Bird Strike in Aviation

Bird Strike in Aviation

Statistics, Analysis and Management

Ahmed F. El-Sayed
Zagazig University
Egypt

Registered Office
John Wiley & Sons, Inc., 111 River Street, Hoboken, NJ 07030, USA
John Wiley & Sons Ltd, The Atrium, Southern Gate, Chichester, West Sussex, PO19 8SQ, UK

Editorial Office
The Atrium, Southern Gate, Chichester, West Sussex, PO19 8SQ, UK

For details of our global editorial offices, customer services, and more information about Wiley products visit us at www.wiley.com.

Wiley also publishes its books in a variety of electronic formats and by print-on-demand. Some content that appears in standard print versions of this book may not be available in other formats.

Library of Congress Cataloging-in-Publication Data

Names: El-Sayed, Ahmed F., author.
Title: Bird strike in aviation : statistics, analysis and management /
   Professor Ahmed F. El-Sayed, Zagazig University, Egypt.
Description: Chichester, West Sussex, UK ; Hoboken, NJ : John Wiley & Sons,
   Ltd, [2019] | Includes bibliographical references and index. |
Identifiers: LCCN 2019005872 (print) | LCCN 2019011286 (ebook) | ISBN
   9781119529828 (Adobe PDF) | ISBN 9781119529798 (ePub) | ISBN 9781119529736
   (hardback)
Subjects: LCSH: Aircraft bird strikes. | Aircraft accidents–Statistics.
Classification: LCC TL553.5 (ebook) | LCC TL553.5 .E37 2019 (print) | DDC
   363.12/482–dc23
LC record available at https://lccn.loc.gov/2019005872

Cover Design: Wiley
Cover Image: ©mokee81/iStock.com

Set in 10/12pt WarnockPro by SPi Global, Chennai, India
Printed and bound in Singapore by Markono Print Media Pte Ltd

10 9 8 7 6 5 4 3 2 1

Contents

Preface

This book provides a comprehensive survey and analysis of the important topic of bird strike in aviation from different points of view including historical, technological, and biological. It is written to appeal to several groups: (i) engineers of many specializations working in airlines, airports, aerospace corporations, and military bases; (ii) technicians in airlines and bases; (iii) pilots and flight crew; (iv) staff of aviation authorities such as the ICAO, EASA, FAA, IATA, etc.; (v) biologists and physicists; (vi) university and high school instructors and students; and (vii) press and media staff.

This book is written in a straightforward way, avoiding the dry and sterile character of many technical books. It attempts to talk to the reader and provide a self-pacing vehicle that enables the reader to obtain a fundamental understanding of bird strike.

Flight has been the dream of mankind ever since they saw birds soaring effortlessly through the sky. Humans have been inspired by the ability of birds to fly and tried to imitate them, although it took until 1903 before they succeeded. However, when humans began to share birds in their airspace collisions started to occur.

A "bird strike" is defined as a collision between an avian creature (bird, bat, or insect) and any kind of aircraft, whether civilian or military and either fixed wing or rotary wing (helicopter). Collisions may be encountered in any flight phase (takeoff, cruise, or landing). Bird strike can be a collision of a single bird or a group (a flock) of birds with an aircraft. Birds also may collide with missiles during their launch. The Space Shuttle *Discovery* hit a vulture during lift-off in 2005.

Looking at one side of the collision, bird strike always leads to the death of the bird or birds. From the other side, the aircraft may experience minor or major damage to its airframe and engines. In the worst case, bird strike may have the catastrophic consequence of complete destruction and the loss of some/all passenger lives. Significant bird strikes that disable engines usually cause complete destruction of the aircraft.

Since the early piston-powered aircraft were noisy and relatively slow, birds usually managed to avoid these aircraft. Consequently, the strikes that happened typically resulted in minor or no damage to windshields, leading edges of wings/tails, or the fuselage. The probability of collision was also small because of the small number of aircraft. Most of the birds learned to stay away from the dangerous airspace in the vicinity of airports.

The onset of the jet age revolutionized air travel, but dramatically magnified the bird/aircraft conflicts. The main reasons are now discussed.

- Wildlife and airports exist near each other.
- There has been a substantial increase in air traffic worldwide. Commercial air traffic increased from about 18 million aircraft movements in 1980 to 38.1 million in 2018 (a 1.8% increase per year) and is expected to reach more than 51 million in 2030. At any instant, there are 35 000 aircraft in air worldwide, some 6000–9000 of which are in the USA.
- Aircraft have assumed a vital role in tactical and logistical military operations.
- The increase in aircraft size, speed, and quietness make it is easier for them to escape the attention of the birds. Commercial air carriers are replacing their older three- or four-engine aircraft fleets with more efficient and quieter, two-engine aircraft. This means there is a reduction in engine redundancy that increases the probability of life-threatening situations resulting from aircraft collisions with flocks of birds.
- A large quantity of air sucked into the modern, powerful jet engines (turbofan and turboprop) and the large dimensions of the engine intakes. Jet engines have also proved to be less resistant than piston engines to collision with the birds.
- There has been a marked increase in the populations of hazardous bird species in many parts of the world in the last few decades. For example, the Canada goose population in the USA and Canada increased at a mean rate of 7.3% per year between 1980 and 2006. Other species showing significant mean annual rates of increase include bald eagles (5.0%), wild turkeys (13.0%), turkey vultures (2.3%), American white pelicans (4.3%), double-crested cormorants (4.9%), and sandhill cranes (4.7%). Billions of birds, bats, and insects use the atmosphere for migration, dispersive movements, and foraging. Environmental protection programs worldwide have contributed to considerable increases in populations of many large-bodied species such as cormorants, cranes, geese, gulls, herons, pelicans, falcons, eagles, owls, vultures, and wild turkeys. As an example, in the USA about 90% of all bird strikes involve species federally protected under the Migratory Bird Treaty Act.

Based on the FAA database, 13 244 bird strikes were reported in 2014, with 581 causing significant damage. These numbers mean that there are 1.5 bird strikes per hour over the whole year. Moreover, about 5000 bird strikes were reported by the USAF in 2010. These statistics have brought attention of many people to the problem of bird strike.

Bird strike is no longer an engineering problem, but rather a social one. The 2016 movie "Sully" with the famous actor, Tom Hanks, is evidence of the social concern about this issue. But do not worry, please fly calm and confident as flying is still the safest method of transportation. Based on a recent study carried out by Northwestern University for the period from 2000 to 2009, in terms of the deaths per one billion passenger miles, flying has the lowest number (just 0.07). The numbers for other modes of transport are bus (0.11), rail (0.43), ferry (3.17), car (7.28); sadly, motorcycle was the worst (212.57).

This important topic encouraged me to write the present book, which is the second book to handle bird strikes worldwide.

The main objective of this book is to review the past and present status of bird strikes, for both civil and military aviation, and to explore how to improve the future, with aircraft and avian creatures sharing the sky.

Many of the topics presented in this book stem from several lectures I have given at different institutions and companies worldwide. These include MIT; the US Air Force

Academy (USAFA), Springfield, Colorado; Embry Riddle Aeronautical University; Boston University; the University of Central Florida; Moscow Institute for Physics and Technology (MIPT); EgyptAir Training Academy; and Zagazig University, Egypt. Conversations with professionals, instructors, and students influenced much of my writing in this book.

Bird strike is part of the global foreign object damage (FOD). FOD is the focus of more than 50% of my research work since the 1970s. Parts of my research was published in my book *Aircraft Propulsion and Gas Turbine Engines* (Taylor & Francis, CRC Press, Boca Raton, FL) in its two editions (2008, 2017).

Chapter 1 provides a historical introduction and many definitions. Bird strike is also known as birdstrike, bird ingestion (for an engine), bird hit and bird aircraft strike hazard (BASH). The distinction between "incident" and "accident" was first made by the ICAO in its annex 13. The term "bird strike" is usually expanded to include other wildlife species, including terrestrial mammals. The term "bird strike" belongs to the larger family called FOD, where the abbreviation FOD stands for both Foreign Object Damage and Foreign Object Debris. "Bird" is a kind of "Debris" while "Strike" leads to "Damage". FOD (foreign object debris) is divided into animate and inanimate sources. Animate sources consist of wildlife, grass, and humans. Wildlife includes ground animates (coyotes, dogs, deer, and snakes) and airborne animates (birds, bats, and insects). Inanimate sources includes broken pieces of concrete, solid stones, tools left by mechanics, pieces of tires, hail, rain, sand, snow, and food remains. More than 90% of FODs can be attributed to avian creatures.

The first bird strike incident was recorded by Orville Wright when his aircraft hit a bird (probably a red-winged blackbird) as he flew over a cornfield near Dayton, Ohio, in 1905. The first fatality was on 3 April 1912, when the aircraft control cable of Calbraith Rodgers struck a gull along the coast of Southern California.

There are three distinct cases for bird strikes; namely: single or multiple large bird(s), relatively small numbers (between 2 and 10 birds) of medium-sized birds, and large flocks of relatively small birds (more than 10 birds).

Bird strikes have caused numerous accidents, resulting in negative impacts on aircraft and airline industry, human casualties as well as harmed wildlife. Annually the International Civil Aviation Organization (ICAO), Federal Aviation Administration (FAA), European Aviation Safety Agency (EASA), Bird Aircraft Strike Hazard (BASH) team, and the International Bird Strike Committee (IBSC) publish statistics for bird strikes.

Many details concerning accidents for civil and military aircraft since 1960 are given. The annual cost for bird strike in the period 1990–2009 for the USA was US\$ 400 million and up to US\$ 2.0 billion for the whole world (estimated by the European Space Agency). The total number of fatal bird strike accidents since 1912 is 55, killing 277 people and destroying 108 aircraft.

Finally, the bird strike committees are listed. The role of these committees in increasing the awareness of the dangers of bird strike as well as managing the risk of bird strikes is summarized.

Chapter 2 discusses bird strike with both fixed and rotary wing aircraft. First, the cases involving both civilian and military fixed-wing aircraft are reviewed. Next, the impact force due to a bird strike is defined. The aircraft locations most impacted by birds are nose and radar dome (radome), windshield and flight cockpit, measuring instruments, landing gear and landing gear system, fuselage, wing, empennage, power plant (engine), and propellers. Details of many accidents and the associated damage in each of the parts are described. A dangerous accident for the military aircraft V-22 Osprey is described.

Then accidents for both civilian and military rotary-wing aircraft (helicopters) are discussed. Accidents encountered by both small and large helicopters are identified. Finally, a brief list of some accidents of both fixed and rotary wing aircraft is tabulated.

Chapter 3 presents statistics for different aspects of bird strike and bird species. Both civilian and military fixed-wing aircraft and rotary wing (helicopters) aircraft are discussed. Only bird strike accidents/incidents having no fatalities are discussed in this chapter, as Chapter 4 handles fatal accidents.

Fixed-wing civil aircraft are reviewed first. Birds are classification based on the critical sites in the aerodrome. Percentages for bird strike with different parts/modules of aircraft powered by turbine-based engines (turbojet and turbofan) or shaft-based engines (piston and turboprop) are reviewed, based on US, European, and Russian statistics.

The majority of bird strikes occur at low altitudes (less than 3000 ft). Empirical relations for the number of reported bird strikes on commercial and general aviation aircraft versus altitude (based on databases of FAA) are presented. Annual and monthly accidents over the last 100 years are discussed. The critical times of day are also identified. Bird strikes for different areas of the globe, including North America, UK, Russia and Asia-Pacific regions are described. Details for dangerous bird species in North America, and their increasing population are also given.

Military aircraft statistics are also discussed; however, fewer publications are available. Catastrophic accidents for different categories of military aircraft (fighter, bomber, etc.) are defined. The critical parts of these aircraft frequently impacted are stated. Annual and monthly statistics for bird strike are given. Statistics for bird strike by flight phase and distance from base are also given.

Next, civil and military helicopters are examined. Annual and monthly statistics for bird strikes with helicopters in the period 1990–2013 are reviewed. The windshields of the helicopter are the most critical part.

Finally, the number of birds killed per year due to collisions with aircraft are identified. The FAA assumes that at least one bird is killed during each bird strike incident. However, in many cases, several birds are killed in each incident. For example, an F16 military aircraft struck a flock of birds, which resulted in 40 dead birds.

Chapter 4 presents, in chronological order, fatal accidents arising from bird strike to both civil and military aircraft. Both fixed wing and rotary wing aircraft are considered. Fatal accidents cover accidents where there are either killed or seriously injured personnel aboard the aircraft or the aircraft itself is severely damaged and cannot be repaired. The first fatal accident was on 3 April 1912, to Calbraith Rodgers. Fatalities in the First Word War were not registered so accidents which were due to birds cannot be differentiated from those due to military actions. A list of fatalities due to bird strike during the Second World War is available and is almost complete. Survey of accidents since the 1950s to the present for civil aircraft are discussed first, followed by military ones and finally civil and military helicopters. The critical phases of flight in both civil and military aircraft are concerned with those near the ground – takeoff, climb, approach, and landing.

Dangerous bird species causing many of the fatal accidents with commercial airplanes, executive jets, and small aircraft in the last 100 years are identified. In Europe, gulls and diurnal raptors pose the greatest danger for both military and civil air traffic. Less dangerous birds are swallows, swifts, pigeons, European starlings, and northern lapwings. Storks, herons, and vultures are dangerous locally. In Europe, most tragic accidents are caused by gulls, European starlings, and northern lapwings. North America (the USA and Canada) have different dangerous bird species.

The aircraft parts struck in fatal accidents are identified. These were engines, windshield, fuselage, radome, headlights, and landing gears. Russian accidents were also reviewed, based on the data available (for the period 1988–1990). Details of about 30 fatal accidents are described. Fatal accidents for military aircraft of 10 countries from the 1950s to 1990s were next summarized. Details of the eight most catastrophic fatal accidents were described. Full details for accidents encountered by aircraft of Royal Norwegian Air Force aircraft during 2016, as an example for a European country, were also given. Finally, the fatal accidents for helicopters in the period 1981–2014 are stated.

Chapter 5 discusses bird migration. Nearly 40% of the world's 10 000 bird species are migratory, with the rest being permanent residents. Birds migrate to move from areas of low or decreasing resources to areas of high or increasing resources. The two primary resources being sought are food and nesting locations. Birds that nest in the northern hemisphere tend to migrate northward in the spring to take advantage of burgeoning insect populations, budding plants, and an abundance of nesting locations. As winter approaches and the availability of insects and other food drops, the birds move south again. Escaping the cold is a motivating factor. However, many species, including hummingbirds, can withstand freezing temperatures as long as an adequate supply of food is available.

The different types of bird migration are north–south, south–north, longitudinal, loop, leap-frog, walk and swim, and short-, medium-, and long-distance migrations.

The four famous major bird migration areas are North America, the Americas, Africa–Eurasia and East Asia–Australasia. North America has four flyways: Atlantic, Mississippi, Central, and Pacific. The Americas has two flyways: North–South America and Alaska. Famous birds flying these different routes as well as their populations are described. Radio telemetry is a technique that is frequently used to determine bird movements over areas ranging from breeding territories of resident bird species to the large distances covered by international migratory species.

Chapter 6 introduces different methods of bird strike management. Management is a vital process to minimize or prevent the losses in lives, aircraft, and money. Many misconceptions are first identified, to select an appropriate procedure for preventing or minimizing bird strikes. The FAA has a National Wildlife Strike Database for Civil Aviation, including all details of bird strike reports since 2000.

There are three aspects to consider for reducing of bird strike hazards: awareness, bird management (control and avoidance), and aircraft design.

Awareness means recognizing the presence, problems, and danger of birds at and around an airport. A thorough identification of different methods for reduction of bird strike hazard is provided.

There are four areas associated with bird management: controlling the airport and its surroundings, air traffic service providers, pilots, and air operators. Managing airport and surroundings rely upon numerous active and passive methods. Active methods include pyrotechnics, bioacoustics, depredation, propane gas cannons, dogs, falconry, shooting, radio-controlled craft, all-terrain vehicles, pulsating lights, dead birds, fake hawks, lasers, chemical repellent, scarecrows, and scaring paints on aircraft. Also, the removal of nests, eggs, and young, as well as sterilization of eggs, are extremely effective methods in reducing bird populations. Passive methods include grass management, managing reforested areas, landscaping, water-habitat management, landfills, managing agricultural programs in air bases. Concerning air traffic service providers, all duties for controllers and flight-service specialists, terminal controllers, tower and ground controllers, as well as flight service specialists are specified. The third task is associated with pilots, who play a vital role during preflight preparation, taxiing, takeoff, climb, en route, approach, landing, and post-flight phases. Finally, the responsibilities of air operators in flight planning and operating principles are described.

Aircraft and aero-engine manufacturers are obliged to fulfill the endurance limits set by aviation authorities (FAA/EASA) in their designs. With rotary wing aircraft, the requested operations for personnel handling helicopters and helipads are identified.

Finally, bird avoidance is controlled using either or both of avian radar and optical devices. Optical devices include high resolution cameras and infrared cameras.

Chapter 7 illustrates bird strike testing for both airframe and engine. Such testing is now a vital step in the certification of any new aircraft and engines. Both American and European authorities issue its certification requirements (FAR and CS) and keep amending them to match any changes in bird ecology and aviation industry. Either real or artificial birds are employed in bird impact test facilities, though real birds are not recommended. A description for bird strike testing with wing and tail fin is thoroughly described. Two case studies (Alenia C-27J Spartan transport aircraft and the recent Russian MS-21 airliner) are discussed.

Bird strike testing for military aircraft is discussed. The canopy and windscreen of the US F35B STOVL as well as the lift fan inlet door in its STOVL mode are discussed. Both FAR and CS-800 Certifications regarding bird strikes with engines are discused. Details of engine test cells for an engine manufacturer are given. Bird strike testing of the turbofan engine powering the Airbus A320 aircraft is described. Post-impact analysis is performed by dismantling the different modules of engine and identifying the damage.

Finally, bird strike tests for helicopters are described. A test case performed at the Southwest Research Institute (SwRI) is described; a 2.2-lb weight bird impacted an inclined surface representing the windscreen at 140 and 180 knots.

Chapter 8 reviews the different numerical methods employed in bird strike simulation. Numerical methods can simulate bird/structure interactions and thus provide the designer with very useful data, such as stress distribution and structural deformation. Numerical methods also enable a parametric study of different materials, different geometry, as well as impact speed (magnitude and direction). They are easier and cheaper compared to experimental bird strike testing.

The finite element method (FEM) has been adopted in most bird strike studies due to its capabilities for handling very complex geometries and providing material behaviors under different loading conditions. MSC/Dytran and LS-Dyna solver codes are frequently used. The three steps employed in numerical methods (pre-processing, solution, and post-processing) are described. Geometry and material of birds are modeled next. Regarding bird geometry, few methods handle a bird's exact shape, so most of them replace the bird shape with either a straight-ended cylinder, a hemispherical-ended cylinder, an ellipsoid, or a sphere. Bird impact behavior on rigid targets are divided into four main stages: initial shock, pressure decay (release), steady state, and pressure termination stage.

Four numerical methods, Lagrangian, Eulerian, arbitrary Lagrangian Eulerian (ALE), and smooth particle hydrodynamics (SPH), are employed. Firstly, these methods are described in detail. Next, these methods are applied to real cases, including both fixed wing and rotary wing (helicopter) aircraft. For fixed-wing aircraft, bird strikes with the leading edges of the wing, and horizontal and vertical tails sections are reviewed. Other case studies, including sidewall structures and windshields, are demonstrated. Finally, impacts on the fan/compressor of its power plants are also reviewed. Switching to helicopters, bird strike with its windshield, rotor, and its spinner are discussed.

The final chapter, Chapter ,9 provides procedures for the identification of bird species involved in bird/aircraft strikes. This is an important part of the overall assessment, management, and wildlife mitigation at airports. The first step in such a process is the collection of remains of bird strike from an aircraft or engine. These remains are sent to appropriate labs, such as the Smithsonian Institution (SI) in the USA or the Central Science Laboratory (CSL) in the UK. Identification methods are either macroscopic or microscopic. Identification by eye is a macroscopic method that depends on the eyes of highly trained personnel. A better result may be achieved by microscopic examination of feathers and blood stains, which can be compared with stored samples. The keratin electrophoresis feather identification process provides much better results when compared with stored feather specimens. DNA analysis of bird remains provides the most accurate method, though it is very expensive. A case study for bird identification using different techniques is described for the crash of Cessna Citation 1 (Model 500) aircraft about 7 km from Wiley Post Airport, Oklahoma, in 2008.

This book is dedicated to my wife, Amany, and my sons, Mohamed, Abdallah and Khalid, for all their love and support. My deepest gratitude goes to my wife for putting up so patiently with the turmoil surrounding a book in progress, waiting until we can breathe a joint sigh of relief at the end of the project.

I would like to thank Mohamed El-Sayed, CEO EA Energy Solutions, for his continuous support and help. Thanks also to Darrell Pepper, University of Nevada Las Vegas, for his continuous fruitful thoughts and discussions. Sincere thanks to Christian Kierulf Aas, Aviation Bird Office, Natural History Museum, University of Oslo, for providing me with valuable information regarding bird strike in Norway.

I am also grateful for the kind assistance of the worldwide bird strike experts and specialists: Mr Anastasios Anagnostopoulos (Head, Wildlife & Biodiversity Management, Environmental Services Department, Athens International Airport), Mr Albert de Hoon (Secretary of World Birdstrike Association, WBA), Dr Marcus Lloyd-Parker (Senior Wildlife Consultant, SafeSky), and Mr W. John Richardson (Director Emeritus & Senior Biologist, LGL Ltd).

I would like to acknowledge the generosity of Michele Guida (University of Naples, Federico II, Italy), for permitting me to use several photos of his results, and of Ahmed Hamed (EgyptAir Company), who shared a lot of information regarding engine performance and bird strike with the EgyptAir fleet.

The sincere support of Tanya Espinosa (USDA-APHIS), Barbara Murphy (National Academies Press), John R Weller (FAA), John A Donald (Robop Limited, UK), Sebastian

Heimbs (Airbus Operations GmbH, Deutschland), Vinayak Walvekar (Engineering Technologies Associates, Inc., USA), Stuart McCallum (University of Sophia, Japan), Marina Selezneva (KU Leuven, Belgium), Sergey K. Ryzhov (Aviation Ornithology Group, Moscow, Russia), Ronald Tukker (Robin Radar), Oliver Walker-Jones (Rolls-Royce), and LtCol Sean Londrigan (US Air Force Academy), is greatly appreciated.

I would also like to thank Teresa Netzler (Manager, Content Enablement & Operations, Knowledge and Learning, Wiley) for her strong support as well as my editor Anne Hunt for her great help and effort in promoting the project since day one.

Ahmed F. El-Sayed

1

Introduction

1.1   Introduction

Flight has been the dream of mankind as we watch birds soar effortlessly through the sky. Humans have been inspired by the ability of birds to fly and have tried to imitate it. Birds first took to the air about 150 million years ago, while humans first began to share their airspace only in 1903. Unfortunately, when aircraft and birds attempt to use the same airspace at the same time, collisions occur.

So, let us first define the term "bird strike". A bird strike is defined as a collision between a fixed- or rotary-wing aircraft during different flight phases and an airborne

Bird Strike in Aviation: Statistics, Analysis and Management, First Edition. Ahmed F. El-Sayed.
© 2019 John Wiley & Sons Ltd. Published 2019 by John Wiley & Sons Ltd.

avian creature (usually a bird or a bat) or a group (a flock) of such avian creatures, resulting in the death or injury to the bird, damage to the aircraft, or both.

As defined by the Federal Aviation Administration (FAA), near-collisions with birds reported by pilots also are considered strikes.

"Birdstrike" (as one word) and "bird hit" are alternatives for the term "bird strike." Bird strike with the aircraft's engine is normally denoted as "bird ingestion."

The hazard associated with bird strike is identified as bird aircraft strike hazard (or BASH). BASH also stands for "birds/wildlife aircraft strike hazard."

Based on the Bird Strike Committee Canada, a bird strike is deemed to have occurred whenever:

- a pilot reports a bird strike;
- aircraft maintenance staff identify damage to any module of aircraft caused by a bird strike;
- ground personnel report seeing an aircraft strike one or more birds;
- bird remains are found on an airside area or within 200 ft of a runway (unless bird's death is due to other reasons).

A possible collision between an aircraft and a flock of birds is illustrated in Figure 1.1. Ground movements of birds may also pose a threat to aircraft, as illustrated in Figure 1.2.

Birds may also collide with missiles during the flight of the missile. Figure 1.3 illustrates a flock of birds (possibly cormorants) surrounding the Space Shuttle *Atlantis* in 2002.

On 26 July 2005, the Space Shuttle *Discovery* hit a turkey vulture [1] during liftoff (Figure 1.4).

Bird strike leads to the death of bird(s). For the flying vehicle, the aircraft or rocket may experience either a minor or major damage of its airframe and/or engines or may have catastrophic destruction with the loss of some/all passenger lives. Complete destruction of aircraft is mostly caused by significant bird ingestion into engines [2].

1.2 Bird Strike: Foreign Object Damage (FOD)

The term "Bird Strike" is usually expanded to include other wildlife species, including terrestrial mammals. Bird strikes cause the greatest amount of FOD (or Foreign Object Damage) of aircraft. (Note that FOD is also the abbreviation for "Foreign Object

Figure 1.1 Birds threatening a flying aircraft [27]. *Source:* An Airbus A330 of China Eastern behind a flock of birds at London Heathrow by NMOS332, B-6543 is Licensed under CC BY-SA 2.0.

Figure 1.2 Birds threatening aircraft on ground movements. *Source:* Courtesy USDA-APHIS [23].

Figure 1.3 Birds are surrounding Space Shuttle Atlantis in 2002. *Source:* Courtesy NASA.

Figure 1.4 A turkey vulture flew right into Space Shuttle Discovery. *Source:* Courtesy NASA.

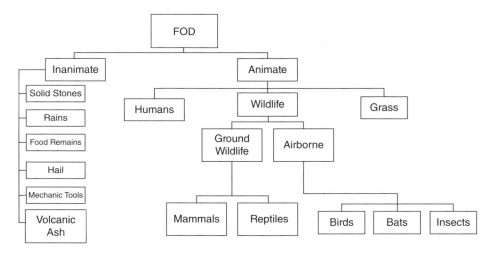

Figure 1.5 Foreign object damage.

Debris.") FOD (Figure 1.5) covers both animate and inanimate sources. "Animates" include wildlife, grass, and humans. Moreover, wildlife includes ground animates (coyotes, dogs, deer, and snakes) and airborne animates (birds, bats, and insects) (Figure 1.6).

(a)

(b)

Figure 1.6 Airborne animates (a) Bird [28]. *Source:* Bald eagle anatomy by Peter K. Burian, Licensed under CC BY-SA 4.0. (b) Bat [29]. *Source:* Flying fox at Royal botanical gardens in Sydney by Hasitha Tudugalle is licensed under CC-BY 2.0.

"Inanimate" includes solid stones, tools left by mechanics, hail, rain, and food remains. Recently, drones have been added to the list of possible FODs.

Drones flying close to airport may threaten aircraft during its takeoff and landing flight phases. On 20 December 2018 flights at Britain's second busiest airport, London Gatwick, were suspended after several sightings of drones flying near the airfield. Disruption was caused to at least 20 000 passengers in the run up to Christmas. Planes were unable to take off and a number of flights scheduled to land were diverted to other airports.

More than 90% of FODs can be attributed to avian creatures [3]. Consequently, a bird strike is one of the most critical FOD. It caused numerous accidents, resulting in aircraft damage and human casualties as well as harmed wildlife.

Bats are the only flying mammals. They generally fly only when it is fully dark or at dusk and dawn. Bats are categorized as either "small" or "large" [4]. Small bats are identified as "insectivorous" bats as they feed exclusively on insects, whereas large bats are called "fruit bats" or "flying foxes" (which are vegetarians). Bats of both types are found everywhere in the world, apart from large (fruit) bats which are not found in the Americas.

Bat strikes by aircraft are relatively rare and form only a low proportion of all wildlife collisions [5].

Small bats navigate using an "on-board radar system" analogous to primary radar and thus rarely collide with aircraft.

Moreover, fruit bats use only trees and caves as roosts and fly at low level to and from their feeding sites of tree-borne fruit. Thus, the risk due to strike of a fruit bat with an aircraft is even lower than for insectivorous bats.

Humans appear on the list of possible FOD constituents, but it is rare that accidents are encountered. One of the saddest accidents occurred on 17 December 2015, when a maintenance crew member for Air India was sucked into a jet engine of the Airbus A-319 aircraft at Mumbai airport. The pilot misinterpreted a signal and switched on the plane's engine, which led to the death of the maintenance crew member.

1.3   A Brief History of Bird Strike

The first known bird strike occurred in Ohio on 7 September 1905 when Orville Wright struck and killed a bird near Dayton [6]. Orville Wright outlined that this strike occurred when his aircraft hit a bird (probably a red-winged blackbird) as he flew over a cornfield near Dayton Ohio (Figure 1.7).

The human first victim of a bird strike was Calbraith Rodgers. (He had made the first transcontinental airplane flight across the USA, traveling from 19 September 1911 to 5 November 1911.) Regretfully, on 3 April 1912 he flew into a flock of birds during an exhibition flight in Long Beach, California. A gull got caught in an aircraft control cable and the plane crashed into the ocean. The pilot's neck was broken and he died a few moments later [6] (Figure 1.8).

The US Navy's first fatality due to a bird strike occurred in 1914 [7], coincidentally the same year it obtained its first aircraft.

It is fortunate that worldwide no fatalities have ever resulted from bat strike with aircraft. However, bat strike incidents in the USA have steadily increased from four in

Figure 1.7 First bird strike in 1905. *Source:* Courtesy Orville Wright Museum.

1990 to 255 in 2014, with a total of 1264 cases in total from 1990 to 2014. No complete similar record is available in Europe, while in Ireland five bat strike incidents have been reported over the 10-year interval, 2006–2015 [5].

In brief, conflicts between birds and aircraft (or bird strike) have increased dramatically in the recent years. The main reasons are:

- A sustained increase in aircraft movements over the last 50 years, with a substantial increase in air traffic worldwide since 1980 [8, 9]. Commercial air traffic increased from about 18 million aircraft movements in 1980 to exceed 28 million in 2007 (at the rate of 1.8% per year). In the USA, passenger enplanements increased from about 310 million in 1980 to 749 million in 2007 (a 3.3% increase per year). Moreover, USA commercial air traffic is anticipated to continue its growth at a rate of 2% per year to exceed 36 million movements by 2020.
- Marked increases in the populations of hazardous bird species in many parts of the world in the last few decades. For example, from 1980 to 2013 the Canada goose population in North America increased from 500 000 in 1980 to 3.8 million in 2013, while the snow goose population increased from about 2.1 million to 6.6 million (according to the US Fish and Wildlife Service). Other species also have escalating growth rates. For example, from 1980 to 2006, the average annual rates of increase are 2.3% for turkey vultures, 4.3% for American white pelicans, 4.7% for sandhill cranes, 4.9% for double-crested cormorants, 5% for bald eagles, and finally 13% for wild turkeys [10].
- Wildlife and airports are close to each other [11].
- Nearly 90% of all bird strikes in the United States are with species protected federally under the Migratory Bird Treaty Act.

Figure 1.8 First bird strike fatality. *Source:* Courtesy National Academies Press [25].

- Birds are less able to detect and avoid modern jet aircraft powered by quieter engines like CFM56, GE90, PW4000, and RR Trent series [12].

Note, however, that commercial aircraft are no longer powered by three or four engines, as in the case of Boeing 707, Airbus A340, and DC10. Most aircraft use two engines that are more efficient and quieter; examples are Airbus A330 and Boeing 777. This reduction in the number of engines decreases the probability of aircraft collisions with flocks of birds.

1.4   Brief Statistics of Bird Strike

There are three distinct cases for bird strikes; namely:

- *single* or multiple large birds;
- *few* numbers of medium-size birds;
- *large* flocks of small mass birds.

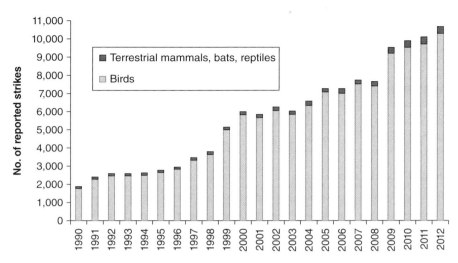

Figure 1.9 Bird strikes for US civilian aircraft. *Source:* Courtesy USDA-APHIS [24].

Bird strikes have negative impacts on the airline industry and represent a real threat to aviation safety. The annual costs of bird strikes exceed one billion US dollars and some collisions have resulted in the loss of human life. The loss in human lives when aircraft crash due to bird strikes proves the need for proper bird management. Each of the International Civil Aviation Organization (ICAO), FAA, European Aviation Safety Agency (EASA), International Bird Strike Committee (IBSC), and Bird Aircraft Strike Hazard (BASH) has a database and publishes annual reports summarizing statistics for these bird strikes. The following brief facts are taken from these databases:

- In the period from 1960 to 2004, bird strikes had catastrophic consequences where more than 120 civilian aircraft have been destroyed and over 250 civilian lives have been lost worldwide. Moreover, some 330 military aircraft were destroyed and over 150 military personnel were killed [13].
- Between 1990 and 2004, US airlines recorded 31 incidents in which pilots had to dump fuel after striking birds on takeoff or when climbing. In each incident, an average of 11 600 gal of jet fuel was dumped.
- In 2012, some 10 000 bird strikes were reported for US civilian aircraft, as displayed in Figure 1.9 [14].
- In 2014, more than 13 343 bird strikes were reported, and some 581 cases were accompanied by significant damage as identified by FAA [14].
- In 2010 some 5000 bird strikes were reported by the US Air Force (USAF).
- The annual cost of civilian aircraft damage due to bird strikes in the USA in the period 1990–2009 was around US$ 400 million.
- Bird strike with airliners may lead to a primary delay and four secondary delays [9]. The aircraft struck by birds is subject to a primary delay or cancelation, while secondary delays or cancelations will be either for subsequent flights to be performed by the airplane or the connecting flights for its passengers. Costs for the primary delay were estimated in 2000 as US$ 75 000. Moreover, costs were US$ 35 000 for a secondary delay and US$ 75 000 for a secondary cancelation [9].

- Costs for annual bird strikes on commercial aircraft worldwide are nearly US$ 1.3 billion [14].

1.5 Classification of Birds Based on Size

Five categories for birds are identified, based on mass. These are listed here, with examples for each category.

1.5.1 Small Birds (Less than 2 lb)

Figure 1.10 illustrates some bird species in this group. Note that small birds threaten small aircraft. Please do not be deceived by their beauty! European starling caused the fatal bird strike accident in aviation history. Eastern Air Lines Flight 375 (a Lockheed L-188 Electra aircraft) crashed on takeoff from Logan International Airport in Boston, MA, on 4 October 1960. Out of the 72 on board people, 62 were killed and 9 had serious injuries.

The pigeon is one of the most famous birds in this group and contributes to many bird strike events. An example of an accident caused by birds of this group is described here. On 2 July 2015, a small flock of gray pigeons (four or five) flew into the Piper PA-38-112 Tomahawk aircraft's path. One bird struck the windscreen, which did not break, but another broke the right-side window, causing bruising to the instructor's shoulder.

Cattle egret, 0.82#

Rock pigeon, 0.81#

Laughing gull, 0.72#

Mourning dove, 0.27#

Killdeer, 0.22#

European starling, 0.19#

House sparrow, 0.062#

Figure 1.10 Small bird species of less than 2 lb. *Source:* Public domain. Guidebook for Addressing Aircraft/Wildlife Hazards at General Aviation Airports, ACRP REPORT 32, National Academies Press [25].

1.5.2 Small–Medium Birds (2–4 lb)

Figure 1.11 illustrates some birds in this group. Among them is the Osprey – sometimes known as the sea hawk – a fish-eating bird of prey. Categorized as a large raptor, it normally reaches 24 in. in length with a 6-ft wingspan. The osprey is not only one of the largest birds of prey in North America, but is also one of the most widespread birds in the world. This large, widely dispersed, bird poses a significant bird strike hazard to all aircraft. Twenty-five osprey bird strikes have been documented at Langley and other US military bases, resulting in approximately US$ 1.3 million in aircraft damage [15].

1.5.3 Medium–Large Birds (4–8 lb)

Figure 1.12 illustrates some kinds of this group. Black vultures are one of the many resident bird species at airports. They pose the biggest risk to aircraft and passenger safety because of their large numbers, large size, and tendency to circle in groups on thermal columns above the airports. Black vultures, which typically weigh between 1.6 and 2.8 kg (3.5–6.1 lb), challenge impact standards for aircraft engines, as engines are designed to best withstand ingestions from birds up to a weight of 1.81 kg (4 lb).

1.5.4 Large Birds (8–12 lb)

Figure 1.13 illustrates some bird species of this group. More than 90% of this group are found in North America. In the last three decades, a significant population increase in this group has been recorded [16].

Among this group is the Canada goose, which is responsible for many accidents. The most famous of them is that frequently identified as "the Hudson miracle." On 15 January 2009 the US Airways Airbus A320 Flight number 1549 struck a flock of Canada geese shortly after takeoff from LaGuardia Airport (New York City) on a flight

Osprey, 3.5#

Mallard, 2.8#

Red-tailed hawk, 2.7#

Herring gull, 2.5#

Figure 1.11 Small – medium bird species (2–4 lb). *Source:* Guidebook for Addressing Aircraft/Wildlife Hazards at General Aviation Airports, ACRP REPORT 32, National Academies Press [25].

Figure 1.12 Medium-large bird species (4–8 lb). *Source:* Guidebook for Addressing Aircraft/Wildlife Hazards at General Aviation Airports, ACRP REPORT 32, National Academies Press [25].

Figure 1.13 Large bird species (8–12 lb). *Source:* Guidebook for Addressing Aircraft/Wildlife Hazards at General Aviation Airports, ACRP REPORT 32, National Academies Press [25].

to Tacoma International Airport (Seattle) [17]. The bird strike caused power loss for both turbofan engines. Captain Sullenberger calmly turned the jet toward the Hudson River and made a water landing, saving the lives of everyone on board. The aircraft was

Figure 1.14 Ditching of Airbus 320 in Hudson River. *Source:* Courtesy NTSB [22].

partly submerged and was gradually sinking when the passengers and crew were rescued (Figure 1.14).

1.5.5 Massive Birds (12–30 lb)

Figure 1.15 illustrates some bird species of this group. The whooping crane represents the lightest in the group (12.9 lb) while the trumpeter swan represents the heaviest (26.2 lb).

Trumpeter Swan, 26.2#   Mute Swan, 26.0#   Wild Turkey, 17.2#

Tundra Swan, 15.9#   American white pelican, 14.0#   Whooping crane, 12.9#

Lappet-faced vulture, 15.4#

Figure 1.15 Massive bird species (12–30 lb). *Source:* Guidebook for Addressing Aircraft/Wildlife Hazards at General Aviation Airports, ACRP REPORT 32, National Academies Press [25].

1.6 Bird Strike Risk

Aircraft damage arising from bird strikes may be significant enough to create a high risk to continue the flight safely. It depends on the type of aircraft, either civilian or military, or fixed/rotary-wing aircraft.

1.6.1 Civilian Aircraft

Damage to airliners arising from bird strikes differs according to the size of aircraft [18]. Small aircraft powered by propellers coupled to either piston engines (examples are the Cessna 172, 182, and 206 and Beechcraft models 18 and 19) or jet engines (examples are the Cessna 208 and Beechcraft model 99) will, when struck by birds, experience structural damage in three main areas – windscreen, control surfaces, and empennage.

The penetration of flight-deck windscreens may lead to the injury of pilots or other persons on board. Figure 1.16 illustrates a windshield of a twin-engine Beechcraft C-99 turboprop that was splattered with blood after striking a western grebe over Arizona in 2009.

When large jet engine aircraft (for example, Boeing B737, 747, 777, 787 and Airbus A320, 380), encounter bird strike they may experience airframe, instruments, landing gear or engine damage. The worst case is bird ingestion into the engines.

Complete engine failure or serious power loss, even if for only one engine, may be critical during the takeoff phase. If birds are ingested into more than one engine then the aircraft will be vulnerable to loss of control. Such a hazardous case arises from either the ingestion of a large flock of medium-sized birds or the ingestion of a smaller number of very large ones. A common consequence of bird ingestion into the engine during the takeoff roll forces the pilot to abort the takeoff.

Figure 1.16 Damage of Beechcraft C-99 turboprop after striking a western grebe.*Source:* Author permission [30].

Figure 1.17 Damage of turbofan engine powering Boeing 757–200 Delta Air Line.*Source:* Courtesy USDA-APHIS [8].

Figure 1.17 illustrates the damage to a turbofan engine powering Delta Air Lines Boeing 757-200 during its takeoff from John F. Kennedy (JFK) International Airport, New York. A flock of birds was ingested into the engine, leading to the damage shown to its fan blades.

The secondary consequence of bird strike (whether ingestion in a large aircraft jet engine or structural impact in a small aircraft) is a partial or complete loss of control.

Bird impacts with the air intake of a static pitot system will result in erroneous readings of flight instruments [19].

Impact with the extended landing gear assemblies in flight may lead to sufficient malfunction of landing gear shock struts, brakes, or nose-gear steering systems. It may cause several kinds of problem during a subsequent landing roll, including directional control problems. The consequences of bird collision with the extended landing gear is illustrated in Figure 1.18, where a Fokker 50 aircraft was forced to land on its belly [20, 31]. The Fokker 50 aircraft, carrying six people, was traveling from Wajir to Nairobi when it struck two marabou storks during takeoff. One of the birds became wedged in a landing gear door, forcing the flight crew to make an emergency belly landing at Jomo Kenyatta International Airport in Nairobi. No injuries were reported.

1.6.2 Military Aircraft

Since military exercises involve flying at high speed and low altitude, aircraft will be exposed to a serious bird strike risk. Consequently, military aircraft are more vulnerable to bird strike accidents than civilian ones. This is usually attributed to a greater proportion of flights conducted at low levels.

Figure 1.19 shows the damage to a C130 Hercules aircraft after hitting a bald eagle near Tacoma, WA, USA. The eagle entered the flight deck through a lower window, close to the pilot's left leg. The pilot was uninjured, but his legs were covered in remains [14].

For such reasons, military pilots of fighter/interceptor aircraft wear helmets with visors to save their life if birds penetrate their aircraft's windscreen.

Figure 1.18 Fokker 50 aircraft landed on its belly at Jomo Kenyatta International Airport, Nairobi, Kenya [20]. *Source:* Photographer permission (G. Banna).

Figure 1.19 Damage of a C130 military aircraft due to Bald eagle strike [32]. *Source:* Reproduced with permission of the author (Ron Rapp).

Figure 1.20 Damage of an F15E military aircraft due to bird strike [14]. *Source:* Reproduced with permission of the photographer (Henrique Oliveira).

As an example of a bird strike with a fighter plane (in 2008), an F-15E was in the final approach to Bagram Airbase, Afghanistan, and was struck by a kite (a giant bird of prey) [14]. Its left engine was completely destroyed, resulting in damage costing US$ 1 million (Figure 1.20).

The US Navy has reported over 16 550 bird strikes on its aircraft, resulting in over 440 aircraft mishaps, 250 damaged enginesdue to FOD, and US$ 372.00 million in damages in the period 1981 to 2011. Additionally, 10 aircraft were destroyed and there was one fatality [5].

1.6.3 Helicopters

Two parts of a helicopter are vulnerable to bird strike – the windshield and the main rotor. In the period 1990 to 2013, bird strikes to civilian and military helicopters resulted in 61 human injuries and 11 fatalities. Repair costs for any incident of bird strike with military helicopters ranges from US$ 12 184 to US$ 337 281.

The number of reported helicopter bird strikes increased from 121 in 2009 to 204 in 2013.

If a bird collides with a helicopter windshield, the bird may penetrate the windshield and knock the pilot unconscious. Pilot incapacitation could cause fatalities for everybody aboard. Figure 1.21 illustrates the case of a bird strike with a helicopter windshield. Birds can also damage the helicopter rotor.

1.7 Severity of Bird Strikes

Impacts with the airframe and engine of an aircraft are estimated by the impact energy. Impact energy depends on the weight of the bird, bird density, the configuration and dimensions of the bird, impact speed, and impact angle. Impact energy is proportional to the mass of bird and the square of the impact speed; or

$$\text{kinetic energy} = (\tfrac{1}{2}\text{mass}) \times (\text{aircraft speed})^2$$

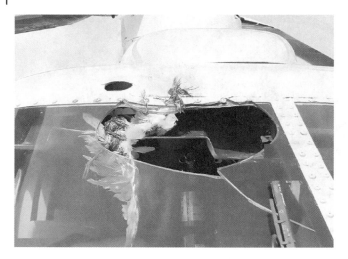

Figure 1.21 Damage of the windshield of a helicopter. *Source:* CREDIT NTSB [26].

For example, a 40% increase in aircraft speed from 250 to 350 KIAS (knots-indicated air speed) results in an 96% increase in impact energy during a bird strike. The speed of the aircraft is more important and more controllable than the size of the bird.

Manufacturers of both airframes and engines employ impact-resistant materials that can reduce bird strike damage. Moreover, the US FAA and the EU JAA (Joint Aviation Authorities) have developed design/certification standards for commercial aircraft. For example, aircraft certification standards for the empennage have been raised to withstand an 8-lb bird impact [21].

1.8   Field Experience of Aircraft Industry and Airlines Regarding Bird Ingestion into Aero Engines

Experience of aero engine designers and operators with bird strikes will be highlighted here [19].

1.8.1   Pratt & Whitney (USA)

1. If large birds impact the fan blades at high speeds, it can cause bends or cusps in their leading edges. However, the engine will continue to operate and the damaged blades can be repaired or replaced easily after landing without removing the power plant [19].
2. The introduction of wide-chord fan blades has reduced bird impact damage.
3. The blade remains contained within the engine and does not damage the aircraft.

1.8.2   General Electric Aviation (USA)

1. Fan blades are robust enough regarding materials, thickness, and overall geometry, to meet bird ingestion criteria.

2. The main innovations in fan blade material (increasing the usage of carbon composite materials together with titanium alloys in the leading edges) lead to increased strength and bird strike resistance.
3. Replacement of damaged fan blade(s) can be done at the flight-line maintenance within a few hours, because of the modular design of today's engines.

1.8.3 Southwest Airlines (USA)

1. Every bird strike is unique, and the extent of damage varies.
2. Bird ingestion into engines may lead to blade damage or clogged pneumatics which may need an internal borescope inspection to assess the damage.
3. The repair process is dependent on where the incident occurred within the airline's maintenance network and the extent of the damage. Thus, getting the right parts and maintenance personnel to the aircraft is the most time-consuming part of the repair process. The aircraft will be back into service within 24–48 hours.
4. Engine removal due to a bird strike is rare and could be needed only once a year.

1.8.4 MTU (Germany)

1. Since turbofan engines have two air streams, namely engine core and bypass, the degree of damage depends on where the strike takes place. If a bird goes through the bypass, the damage is likely to be less severe. However, if the bird enters the engine core, then the bird strike may cause damage to the low-pressure compressor/booster and high-pressure compressor.
2. Damage to compressor blades may lead not only to their failure but also cause severe damage to the entire engine.
3. Bird remains also may enter the turbine and the internal air system, leading to blockages of the engine-cooling flow.

1.8.5 FL Technics (Vilnius, Lithuania)

1. Engine removal – post-bird strike – is required, on average, in up to 5% of cases. Normally the damaged parts, such as the fan blades, are replaced or structural repairs are carried out on the engine inlet.
2. Turbofan engine design allows birds to be directed to the bypass duct, where less damage can be encountered.
3. Usually, single bird ingestion does not create a major problem.

1.9 Bird Strike Committees

The following bird strike committees are the most famous ones:

- *Worldwide*. World Birdstrike Association – http://worldbirdstrike.com. A previous very active committee was the International Bird Strike Committee (IBSC).
- *USA*. Bird Strike Committee – www.birdstrike.org
- *Canada*. Bird Strike Association of Canada – www.canadianbirdstrike.ca

- *Italy*. Bird Strike Committee – www.enac.gov.it/La_Regolazione_per_la_Sicurezza/ Infrastrutture_Aeroportuali/Wildlife_strike/index.html
- *Germany*. Bird Strike Committee – http://www.davvl.de/de
- *Central America*. Regional Committee of Central America, the Caribbean and South America for the Prevention of Bird and Wildlife Hazards (CARSAMPAF) – www .comitecarsampaf.org
- *Australia*. Australian Aviation Wildlife Hazard Group – www.aawhg.org

The roles of existing bird strike committees are as follows:

- exchange best practice information for bird strike management programs and techniques;
- an increase of the awareness of the bird strike threat to aircraft operational safety;
- explore new technologies to minimize the risk of bird strikes;
- enhance bird strike reporting;
- formulate global strategies to address wildlife strike hazards;
- provide an opportunity for networking, collaboration, and coordination between countries, airlines, and other stakeholders.

References

1 Nagy, A. (2013). A brief history of animals and rocket launches not getting along. https://gizmodo.com/a-brief-history-of-animals-and-rocket-launches-not-gett-1301141640 (accessed 27 December 2018).
2 Blokpoel, H. (1976). *Birds Hazards to Aircrafts: Problems and Prevention of Bird/Aircraft Collisions*. Clarke, Irwin & Company.
3 Mao, R.H., Meguid, S.A., and Ng, T.Y. (2008). Transient three dimensional finite element analysis of a bird striking a fan blade. *International Journal of Mechanics and Materials in Design* 4 (1): 79–96.
4 Non Avian Wildlife Hazards to Aircraft. https://www.skybrary.aero/index.php/Non_Avian_Wildlife_Hazards_to_Aircraft (accessed 27 December 2018).
5 Kelly, T.C., Sleeman, D.P., Coughlan, N.E. et al. (2017). Bat collisions with civil aircraft in the Republic of Ireland over a decade suggest a negligible impact on aviation safety. *European Journal of Wildlife Research* 63: 23.
6 Dukiya, J.J. and Gahlot, V. (2013). An evaluation of the effect of bird strikes on flight safety operations at international airport. *International Journal for Traffic and Transport Engineering* 3 (1): 16–33.
7 Department of the Navy (2011). Navy Bird/Animal Aircraft Strike Hazard Program Implementing Guidance. http://www.public.navy.mil/NAVSAFECEN/Documents/aviation/operations/NAVY_BASH_PROG_GUIDANCE.pdf (accessed 27 December 2018).
8 Cleary, E.C. and Dolbeer, R.A. (2005). *Wildlife Hazard Management at Airports: A Manual for Airport Personnel*, 2e, 1–135. http://www.fwspubs.org/doi/suppl/10.3996/022017-JFWM-019/suppl_file/10.3996022017-jfwm-019.s6.pdf (accessed 27 December 2018).
9 Allan, J.R. (2000). The costs of bird strikes and bird strike prevention. In: *Human Conflicts with Wildlife: Economic Considerations*. Paper 18.

10 Dolbeer, R.A. and Wright, S.E. (2008). Wildlife Strikes to Civil Aircraft in the United States 1990–2007, Federal Aviation Administration National Wildlife Strike Database Serial Report Number 14. http://digitalcommons.unl.edu/cgi/viewcontent.cgi?article=1023&context=birdstrikeother (accessed 27 December 2018).

11 Uhlfelder, E. (2013). Bloody skies: the fight to reduce deadly bird-plane collisions. *National Geographic* http://news.nationalgeographic.com/news/2013/10/131108-aircraft-bird-strikes-faa-radar-science (accessed 27 December 2018).

12 Kelly, T.C., Bolger, R., and O'Callaghan, M.J.A. (1999). The behavioral response of birds to commercial aircraft. *Bird Strike '99, Proceedings of Bird Strike Committee-USA/Canada Meeting*, Vancouver, B.C., Canada: Transport Canada, Ottawa, Ontario, Canada, pp. 77–82.

13 Shobakin, H.O. (2009). Bird Prevention: Control techniques adopted by FAAN. A paper presented at the Ramp Safety Event, Mallam Aminu Kano International Airport, Kano, Nigeria, pp. 23–35.

14 Peltier, J. (2016). My Scariest Bird Strike. https://disciplesofflight.com/scariest-bird-strike (accessed 27 December 2018).

15 Washburn, B., Olexa, T. and Dorr, B. (2008). Assessing Bird-Aircraft Strike Hazard (BASH) Risk Associated with Breeding and Migrating Osprey, Bird Strike Committee Proceedings, Bird Strike Committee USA/Canada, 10th Annual Meeting, Orlando, Florida. http://digitalcommons.unl.edu/birdstrike2008/47/ (accessed 27 December 2018).

16 Dolbeer, R.A. and Eschenfelder, P. (2003). Amplified bird-strike risks related to population increases of large birds in North America. Proceedings International Bird Strike Committee 26, Volume 1: 49–67.

17 US Airways Airbus A320 Flt 1549 Hudson River Landing FS2004. https://www.youtube.com/watch?v=CE1stHaC7AA (accessed 27 December 2018).

18 Eschenfelder, P.F. (2015). High Speed Flight at Low Altitude: Hazard to Commercial Aviation? International Bird Strike Committee, USA/Canada 7th Annual Meeting, Vancouver, BC. http://digitalcommons.unl.edu/birdstrike2005/4/ (accessed 27 December 2018).

19 Seidenman, P. and Spanovich, D. (2016). How Bird Strikes Impact Engines. http://www.mro-network.com/maintenance-repair-overhaul/how-bird- strikes-impact-engines (accessed 27 December 2018).

20 Kitching, C. (2015). Small plane forced to land on runway without landing gear after dead bird becomes wedged in door. www.dailymail.co.uk/travel/travel_news/article-2897126/Skyward-plane-makes-belly-landing-Nairobi-s-Jomo-Kenyatta-International-Airport.html (accessed 27 December 2018).

21 O'Callaghan, J. Bird-Strike Certification Standards and Damage Mitigation, National Transportation Safety Board. https://www.ntsb.gov/news/events/documents/oklahoma_city_ok-2_web_bird_strike_cert_and_damage_john_ocallaghan.pdf (accessed 14 January 2019).

22 National Transportation Safety Board (2010). Loss of Thrust in Both Engines After Encountering a Flock of Birds and Subsequent Ditching on the Hudson River US Airways Flight 1549 Airbus A320-214, N106US Weehawken, New Jersey. https://www.ntsb.gov/news/events/Pages/Loss_of_Thrust_in_Both_Engines_After_Encountering_a_Flock_of_Birds_and_Subsequent_Ditching_on_the_Hudson_River_US_Airways.aspx (accessed 27 December 2018).

23 Dolbeer, R.A., Wright, S.E., Weller, J. and Begier, M.J. (2012). Wildlife Strikes to Civil Aircraft in the United States, 1990–2010. Serial Report Number 17. FAA and USDA-APHIS.

24 Dolbeer, R.A., Wright, S.E., Weller, J. and Begier, M.J. (2013). Wildlife Strikes to Civil Aircraft in the United States, 1990–2012. Serial Report Number 19. FAA and USDA-APHIS.

25 Guidebook for Addressing Aircraft/Wildlife Hazards at General Aviation Airports. https://doi.org/10.17226/22949 (accessed 27 December 2018).

26 Aerossurance (2017). Safety Lessons from a Fatal Helicopter Bird Strike http://aerossurance.com/helicopters/fatal-s76c-birdstrike-2009 (accessed 27 December 2018).

27 An Airbus A330 of China Eastern behind a flock of birds at London Heathrow. https://en.wikipedia.org/wiki/Bird_strike#/media/File:B-6543_(7788321152).jpg (accessed 27 December 2018).

28 Bald Eagle flying over ice. https://upload.wikimedia.org/wikipedia/commons/e/e3/Bald_Eagle_flying_over_ice_%28Southern_Ontario%2C_Canada%29.jpg (accessed 27 December 2018).

29 Flying fox. https://upload.wikimedia.org/wikipedia/commons/c/ca/Flying_fox_at_botanical_gardens_in_Sydney_%28cropped%29.jpg (accessed 27 December 2018).

30 Curtis, T. (2009). Full Size Photos from the Show Low Bird Strike Now Available. http://www.birdstrikenews.com/2009/12/full-size-photos-from-show-low-bird.html (accessed 27 December 2018).

31 Plane Fokker 50 Crashes Lands at JKIA. https://www.youtube.com/watch?v=vZV0rSrHQFs (accessed 27 December 2018).

32 Rapp, I. (2005). C-130 Bird Strike. https://www.rapp.org/archives/2005/08/c130_bird_strike.

2

Aircraft Damage

CHAPTER MENU

2.1 Introduction

Birds and aircraft occupy the same airspace and collisions between the two are inevitable. Early piston-powered aircraft (1903–1940) were noisy and relatively slow. Birds could usually avoid these aircraft, and typically bird strikes resulted in minor or no damage to windshield, leading edges of wings and horizontal/vertical tails, or the fuselage.

The probability of collision was also small because of the low numbers of aircraft. Moreover, most birds quickly became acquainted with the noise of aircraft propellers and learned to stay away from the dangerous areas around airports.

Bird Strike in Aviation: Statistics, Analysis and Management, First Edition. Ahmed F. El-Sayed.
© 2019 John Wiley & Sons Ltd. Published 2019 by John Wiley & Sons Ltd.

Figure 2.1 Birds threat to aircraft.

The onset of the jet age revolutionized air travel but dramatically increased bird–aircraft conflicts (Figure 2.1). In brief, conflicts between birds and aircraft (known as bird strike) have increased dramatically in the recent years. Such an increase in bird strike is due to several reasons. Some of these reasons have been mentioned in Section 1.3, and are summarized as:

- a sustained increase in aircraft movements over the last 50 years, with a substantial increase in air traffic worldwide since 1980;
- marked increases in the populations of hazardous bird species [28] in many parts of the world in the last few decades, which is a direct conjugate to the worldwide environmental protection programs;
- wildlife and airports are close to each other;
- most aircraft use more efficient and quieter engines.

Note, however, that most aircraft use just two engines, which decreases the probability of aircraft collisions with flocks of birds.

The problem has been further aggravated by the following additional reasons:

- aircraft have had a vital role in tactical and logistical military operations;
- low altitude and high-speed penetration (approaching or exceeding the speed of sound) place military aircraft in areas of high bird density, which increase the probability of serious aircraft damage;
- the increase in both of aircraft size and speed together with their quietness makes it easier for them to escape the attention of the birds [1–3];
- the great air mass flow rate ingested into the modern, powerful jet engines (examples are GE90 turbofan and A400 turboprop engines) and the large diameters of the engine intake [4, 5];
- rotating modules of jet engines are less resistant to bird strike than piston engines;
- billions of birds, bats, and insects use the air for their seasonal migration, dispersive movements, and foraging.

The bird strike threat is not only limited to large aircraft but also includes smaller general aviation (GA) category aircraft (both fixed and rotary wing). Since most of the bird strikes are experienced near to the ground, the threat may be higher for such aircraft. Air taxi, commuter, and other GA aircraft operate from smaller aerodromes where bird strike management is little or nonexistent. The consequences of a bird strike on GA aircraft may also be severe. This can be due to many reasons, including single pilot operation and a single engine, its light airframe structure, and lack of redundancy in control systems. Finally, there are no specific bird strike certification requirements for GA category aircraft except for windshields of commuter aircraft.

2.2 Accidents vs. Incidents

A bird strike may result in minor or major damage to the aircraft and may lead to injuries and fatalities. The costs of aircraft repair range from several thousand to several millions of dollars. Pilots always worry about bird strike with their planes but, fortunately, passengers rarely think about the issue.

A bird strike may be identified as either an incident or an accident. The International Civil Aviation Organization (ICAO) identified both in its Annex 13 [6].

2.2.1 Accident

An accident is an occurrence anytime between the boarding and disembarking of an aircraft [6], in which:

A. A *person* is fatally or seriously injured because of:

- being in the aircraft, or in contact with any part of the aircraft, or
- exposed to jet blast.

These injuries do not include natural causes or infliction by other persons, or if the injuries are stowaways hiding outside aircraft cockpit

B. The *aircraft* sustains damage or structural failure which:

- negatively affects the structural strength or aircraft flight performance
- would normally require major repair or replacement of the affected component.

C. The *aircraft* is *missing* or is completely inaccessible.

Two important notes are listed in Annex 13 [6] and are reproduced here:

Note 1. For statistical uniformity only, an injury resulting in death within 30 days of the date of the accident is classified as a fatal injury by the ICAO.
Note 2. An aircraft is considered to be missing when the official search has been ended, and the wreckage has not been found.

2.2.2 Serious Injury

As defined in Annex 13 [6],

> *serious injury is an injury which is sustained by a person in an accident and which:*
>
> *(a) requires hospitalization for more than 48 hours, commencing within seven days from the date the injury was received; or*

(b) *results in a fracture of any bone (except simple fractures of fingers, toes, or nose); or*

(c) *involves lacerations which cause severe hemorrhage, nerve, muscle, or tendon damage; or*

(d) *involves injury to any internal organ; or*

(e) *involves second or third-degree burns, or any burns affecting more than 5% of the body surface; or*

(f) *involves verified exposure to infectious substances or injurious radiation.*

2.2.3 Incident

An incident is defined as

> *an occurrence, other than an accident, associated with the operation of an aircraft which affects or could affect the safety of operation.*

The types of aircraft incidents described below are of main interest to the ICAO for its accident prevention studies [7].

(a) *Engine failure. Failures of more than one engine on the same aircraft and failures which are not confined to the engine, excluding compressor blade and turbine bucket failures.*

(b) *Fires. Fires which occur in flight, including those engine fires which are not contained in the engine.*

(c) *Terrain and obstacle clearance incidents. Occurrences which result in danger of collision or actual collision with terrain or obstacles.*

(d) *Flight control and stability problems. Occurrences which have caused difficulties in controlling the aircraft, e.g. aircraft system failures, weather phenomena, operation outside the approved flight envelope.*

(e) *Takeoff and landing incidents. Incidents such as undershooting, overrunning, running off the side of runways, wheels-up landing.*

(f) *Flight crew incapacitation. The inability of any required flight crew member to perform prescribed flight duties as a result of reduced medical fitness. Decompression is resulting in an emergency descent.*

(g) *Near collisions and other air traffic incidents. Near collisions and other hazardous air traffic incidents including faulty procedures or equipment failures.*

2.3 Consequences of Bird Strike

Birds usually collide with aircraft engines as well as other different parts of the aircraft structure (Figure 2.2) including the nose cone, radome, instrumentations, windshield, wings, and empennage. The forward-facing edges of an aircraft, especially the wings, nose cone, and systems, are most crucial. Also, birds can smash a hole and disrupt the air pressure inside the aircraft.

Aircraft systems are vulnerable to bird strike. External instruments such as air data sensors, lights, antennae, anti- and de-icing systems as well as undercarriage modules are vulnerable to direct bird strike.

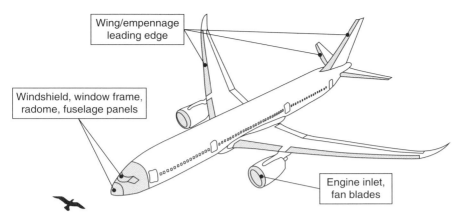

Figure 2.2 Locations of bird collisions with aircraft structure [41].

The canopy, radomes, fuselage, and tail units of military aircraft are subjected to bird strikes. Sometimes bird strikes result in cracking of the surface of the canopy or windshield, which may lead to crew injuries (including fatalities) and damage to flight controls and instrumentation. These cracks can sometimes influence the air pressure inside the cabin and result in altitude loss or other flight-related problems.

The severity of the damage done by the bird strike depends on a few factors, including speed and direction of impact, as well as the weight and size of the bird. Significant changes in materials and technology are required to minimize this damage. Composite materials are now extensively used in both the airframe and engines of aircraft.

In some cases, birds are ingested into the core of jet engines or collide with the propellers of turboprop or piston engines [8]. If sucked into the engine, the bird can disrupt the rotary motion of the fan/compressor blades, resulting in engine shut down. Two-engine planes can fly on a single engine and can glide if both are shut down. Skilled pilots can put the plane down somewhere safely, although the passengers may feel uncomfortable during the pilot's corrective actions [9]. Bird impact with propellers is less dangerous than impact with turbojet/turbofan engines.

Flocks of birds are even more dangerous, as they can have a much nastier impact collectively. It may be concluded that bird strike is a significant threat to flight safety and has caused a few accidents resulting in human casualties. As stated by Thorpe [10], there is one bird strike accident resulting in human death per billion (10^9) flying hours for civil aircraft. Moreover, some 65% of bird strikes cause little damage to the aircraft [11]. However, these collisions are usually fatal to the impacting bird(s).

Analysis of bird strike data for helicopters has only recently gained interest. As stated in Appendix (A) of Advisory Circular No. 29-2C [33], the number of reported bird strikes with helicopters in the period 1990–2005 reached 370 cases out of the 64734 reported bird strikes to civil aircraft. The lack of reporting gives such a low percentage (0.6%) [47]. The outcome of recent interest in reporting bird strikes with helicopter is seen in 665 cases reported in the USA in the years 2015–2017, based on records of the USDA Wildlife Services. This gives an annual strike count of just over 200, a figure consistent with the average US strike count since 2010 [47]. Over 50% of the bird strikes to helicopters, and the damaging strikes, occurred during the en-route phase of flight

(101–500 ft AGL). The climb and approach phases, respectively, are the next critical flight phases [33].

The most frequently impacted modules in a helicopter, arranged in descending order, are windshield, rotor, nose, fuselage, engine, radome, tail, landing gear, and lights [33].

The most dangerous birds are gulls, waterfowl, vultures, and raptors [33].

2.4 Impact Force

A common misconception is that birds are soft and spongy. After all, you may think, they are made of big fluffy feathers, the same thing our pillows are made of. Actually, a bird is a big ball of flying flesh and bones with a razor sharp point at the front. Bird strike results in a huge impact force that can cause substantial damage to aircraft components and may even lead to a crash.

The consequences of a collision between a bird and an aircraft (shown in Figure 2.3) depend on the following variables:

- the bird's mass m:
- the velocity of collision v; and
- the collision time Δt.

Since the bird's velocity is negligible, the collision velocity is taken to be the velocity of the aircraft.

Thus, the impact force F due to bird strike can be described by the equation:

$$mv = F\Delta t \tag{2.1}$$

The resultant force F is expressed by:

$$F = \frac{mv}{\Delta t} \tag{2.2}$$

The collision time is defined in several ways. It is assumed to be the time of transit of the bird's length (which is taken to be 1 ft, 0.3048 m) [12], thus if the speed is assumed to be 100 m/s, then $\Delta t = 0.003$ second. Alternatively, collision time is given a fixed value, $\Delta t = 0.03$ second [13]. These are the upper and lower limits as identified by the industry.

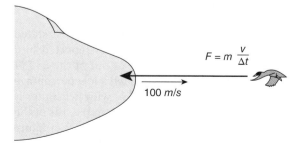

$$F = m\,\frac{v}{\Delta t}$$

100 *m/s*

Figure 2.3 Impact of a bird with aircraft.

The bird's mass m ranges from tens of grams to tens of kilograms. For example, a goose weighs about 3 kg, a stork up to 5 kg, and a pelican can be 10 kg or more.

For jet aircraft at takeoff, approach, and landing phases of a flight, the aircraft velocity v can be as much as 100 m/s (223.7 mph). Substitute this value of v into Eq. (2.2) with $m = 10$ kg, $v = 100$ m/s, and $\Delta t = 0.03$ second, the resultant force will be:

$$F = \frac{10 \times 100}{0.03} = 33.3\,\text{kN}$$

Alternatively, if the time of collision is assumed 0.003 seconds, then the impact force will be

$$F = 333\,\text{kN}$$

Consider the force ($F = 333$ kN) to be spread over an area the frontal size of the bird. This area is approximately width (0.4 m) × height (0.1 m). Thus, the bird area $A = 0.04\,\text{m}^2$.

The corresponding pressure is:

$$p = \frac{F}{A} = \frac{333000}{0.04} = 8.325\,\text{MPa}$$

The aircraft structure either resists this pressure or fails and causes structural damage to the aircraft.

Moreover, since most birds are flying in flocks, the force calculated above must be multiplied by the birds in the flock that hit the aircraft. As a result, such an interaction can have a serious impact on the safety of a flight. An example of bird strike damage can be seen in Figure 2.4.

Finally, both bird size and the location of the collision on the structure of the aircraft are significant factors in a bird strike. The Bird Strike Committee (USA) outlined that the impact force of a 5.4 kg Canadian goose hitting a plane flying at 150 mph (241 kph) will be equivalent to the force generated on the ground by a mass of 454 kg dropped from a height of 3 m.

Figure 2.4 Damage due to bird strike. *Source:* courtesy NTSB [37].

2.5   Locations of Bird Strike Damage for Airliners

All the time thousands of aircraft share air space with millions of birds. Such coexistence yields several tens of bird strikes (30–50) with aircraft every day. This section discusses locations of bird strikes on aircraft. Several accidents will also be discussed. The accidents discussed in this chapter are those without any fatalities but may include some injuries. However, all the accidents will include structural damage, and the repair may cost several million dollars. A survey for these accidents is described in fine detail in an FAA report [14].

Detailed discussion follows for the aircraft modules struck by birds, namely nose and radome, windshield, flight cockpit, landing gears and systems, fuselage, wings, empennage, power plants, and propellers.

2.5.1   Nose and Radar Dome (Radome)

Figure 2.5 illustrates the impact of a bird on the nose cone of a C-130.

Figure 2.6 illustrates a CRJ-200 after a goose struck its radome at 250 knots. There was major damage to the radome and its electronic devices.

Figure 2.7 illustrates the damage to the nose of a Fed Ex MD11 cargo aircraft due to a bird strike on 1 March 2012. At nearly 9000 ft, during a climb, the plane encountered a flock of nearly 20 birds.

Figure 2.8 illustrates the large dent and hole underneath the windshield of a United Airline's 737-900 aircraft. The aircraft landed safely and there were no injuries among the 151 passengers on board.

In March 2014, an Airbus A320 struck a herring gull during a climb at 300 ft above ground level (AGL). The radome suffered the damage shown in Figure 2.9 and blood was spattered over the left windshield [15]. The pilot managed to land safely in a nearby airport. The aircraft was out of service for 7 hours while the radome was replaced.

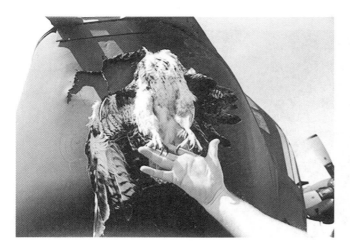

Figure 2.5   Damage to C-130 due to Hawk impact on the nosecone [39]. *Source:* Public Domain.

Figure 2.6 CRJ-200R with damaged radome. *Source:* courtesy USDA-APHIS [3].

Figure 2.7 Damage to the nose of Fed Ex MD11 cargo aircraft. *Source:* courtesy USDA-APHIS [15].

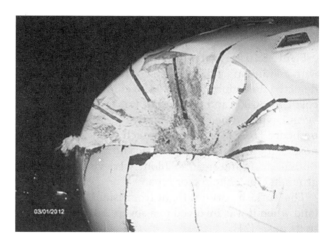

Figure 2.8 Damage to the nose cone of United Airlines B737 aircraft. *Source:* courtesy USDA-APHIS [15].

Figure 2.9 Damage to radome of Airbus 320. *Source:* Courtesy USDA-APHIS [15].

Figure 2.10 illustrates the large dent in a radome of a Boeing 757 [15] due to bird strike with a Franklin's gull during a climb (4600 ft AGL) in September 2014. The flight crew detected no abnormalities and so continued to the destination airport. Aircraft repairs cost US$ 30 000 and the aircraft was out of service for 24 hours.

An Egyptair Boeing 737-800 flying from Cairo on 11 March 2016 with 71 passengers on board [16, 17] was on approach to the runway in London when a bird hit it on the nose of the aircraft (Figure 2.11). The radome cone was penetrated, leaving a huge hole. Both feathers and smeared blood were visible over the front of the plane. The plane was grounded for 21 hours for repair.

The Turkish Airlines flight from Istanbul to Nevşehir in Turkey was on approach to land when it was involved in a severe hit with a unlucky bird, causing a severe damage to its nose [21].

Figure 2.10 A large dent in a radome of a Boeing 757. *Source:* courtesy USDA-APHIS [15].

(a)   (b)

Figure 2.11 A huge hole in a radome of an Egyptair Boeing 737-800: (a) aircraft, (b) nose [16]. *Source:* Reproduced with permission from EgyptAir.

Figure 2.12 Rupture hole in the radome of Russian aircraft An-24. *Source:* courtesy Aviation Ornithology Group, Moscow, Russia [18].

There were 125 passengers on board that Boeing 737–800, and there were not believed to be any injuries suffered on the landing on Tuesday.

Figure 2.12 illustrates a ruptured hole in the radome of Russian aircraft An-24 after a bird strike with a rock pigeon and some other large bird [18].

2.5.2 Windshield and Flight Cockpit

During a horizontal level cruise at constant speed, an aircraft is subjected to four forces. These are two opposite and equal vertical forces (lift and weight) and two opposite and equal horizontal forces (thrust and drag). An aircraft cockpit windshield is inclined to the horizontal axis (taken as parallel to the aircraft centerline). Consequently, the windshield carries two components of surface loads, a backwardly directed, horizontal surface drag force and an upwardly directed, vertical surface lift force. In most cases, bird strike is an additional horizontal force.

Most front windshields are made of several layers (usually three) of mineral glass for scratch resistance and clear plastic (or glass) laminated together. They are about 1–3 in thick (depending on the type of aircraft) and very heavy (20–100 lb). The outer layer is mainly for foreign object damage (FOD) resistance. The windshield acts as a barrier and aircraft can be flown even if it suffers damage, although pilots are understandably loath to do so [3]. The middle and inner layers are the load carrying layers. The side windows are made of a plastic polymer and are not quite as thick.

As per Federal Aviation Administration (FAA) additional requirements (14 CFR Part 25-775), the aircraft must be able to continue in flight and safely land after the windshield and windows have been impacted by a 1.8 kg bird at cruise speed (V_c) at mean sea level.

Figure 2.13 illustrates typical accidents due to bird strike with aircraft windshields.

Figure 2.13 Typical accidents due to bird strike with aircraft windshields. *Source:* courtesy USDA-APHIS [3, 15].

Figure 2.14 Damage to windshield of PA-34 aircraft. *Source:* courtesy USDA-APHIS [3].

Figure 2.14 displays the damaged windshield of a PA-34 aircraft after striking a pair of red-breasted mergansers during the plane's approach phase (800 ft AGL). The birds penetrated both windshields, and the pilot was luckily unhurt [3].

Figure 2.15 shows damage to the windshields of a Cessna 206 aircraft struck by a single vulture on takeoff. The bird penetrated the windshield, severely injuring the pilot [3].

In many cases when birds penetrate the aircraft's windshield, the flight deck is spoiled by birds' feather, blood, and remains, as shown in Figure 2.16.

Another example of such an accident is illustrated in Figure 2.17. In November 2009 this Beech C-99 aircraft had started its descent at an altitude of 11 000 ft above mean sea level (msl). A bird impacted the upper part of the captain's windshield, resulting in a football-sized hole [35]. Blood, tissue matter, and windshield fragments came into the

Figure 2.15 Damage to windshield of Cessna 206 aircraft. *Source:* courtesy USDA-APHIS [3].

Figure 2.16 Typical accidents due to bird strike penetrating windshields leaving remains in flight cockpit. *Source:* courtesy NTSB [42].

Figure 2.17 Windshield and flight deck accident for Beech C-99 aircraft. *Source:* Reproduced with permission from Todd Curtis [43].

cockpit. The captain managed to land safely, though he suffered facial lacerations, bruising, and some lacerations on his chest.

2.5.3 Landing Gear and Landing Gear Systems

Extended landing gear assemblies are subject to bird strike during takeoff/climb and descent/landing phases, as illustrated in Figure 2.18. Such collisions may lead to malfunction of brakes or nose gear steering systems, which in turn can cause problems in directional control during subsequent normal or emergency landing. In some cases, a bird strike on the takeoff roll forces pilots to abort takeoff. If this decision is made after the V_1 speed or if there is a delayed or incomplete brake response then the aircraft may encounter a runway excursion.

Figure 2.19 illustrates a peregrine falcon wedged in the landing gear of an Embraer 190 aircraft during landing roll in August 2014. Fortunately, there was no damage to aircraft. From 1990 to 2014, 274 cases were recorded for peregrine falcons striking civil aircraft in the USA [15].

An example for the effect of a bird strike on tires is shown in Figure 2.20. A Boeing 757-200 suffered a bird strike during its takeoff roll at Amsterdam in July 2009. The pilot aborted takeoff and applied maximum brakes. Brake heating caused the tires to heat up. A fail-safe design is to incorporate a plug in the tires which heats upon applying maximum braking; this allows the tires to deflate and break up instead of exploding. This plug resulted in deflated tires which, upon rolling, caused the tire walls to separate.

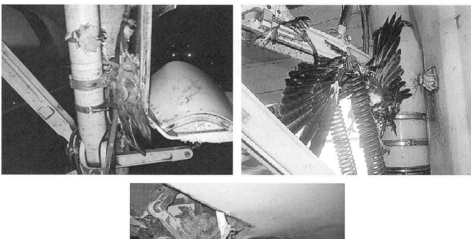

Figure 2.18 Landing gear accidents due to bird strike. *Source:* courtesy USDA-APHIS [15].

Figure 2.19 A peregrine falcon wedged in the landing gear of an Embraer 190. *Source:* courtesy USDA-APHIS [15].

Figure 2.20 Damaged tires of Boeing 757-200 after an aborted takeoff due to bird strike. *Source:* courtesy USDA-APHIS [15].

Figure 2.21 illustrates a Jak-40 (produced in the former Soviet Union) being dragged away from the landing strip after a bird strike [18].

The Boeing company [20] suggested the following procedure if a bird strike is confirmed or suspected when the landing gear is extended during takeoff or landing with high lift deployed. The pilot must:

- *use available system information to assess possible damage to flight controls and high lift devices, and make minimal possible changes in aircraft configuration appropriate to the flight phase;*
- *use available system information to assess possible damage to landing gear and associated systems, including exposed electrical, pneumatic, and hydraulic systems, and potential effects on the ability to steer and stop on the runway.*

Figure 2.21 Russian aircraft Jak-40 experiencing a bird strike. *Source:* courtesy Aviation Ornithology Group, Moscow, Russia [18].

2.5.4 Fuselage

In Sep 2012, an anhinga bird stuck a Pilatus PC-12/45 during its climb phase (6000 ft AGL) out of Florida airport (Figure 2.22). The bird penetrated the fuselage [36].

Figure 2.22 A bird penetrated the fuselage of Pilatus PC-12/45. *Source:* courtesy USDA-APHIS [36].

Figure 2.23 F-16 canopy after a bird strike [44].

Figure 2.23 illustrates the canopy of an F-16 after a bird strike.

2.5.5 Wings

The FAA additional regulations (under 14 CFR Part 25-571) state that an aircraft must be able to continue its flight and land safely after its wing structure has been impacted by a 1.8 kg (4 lb) bird at cruise speed (V_c) at mean sea level [23]. Typical damage to wings due to bird strikes are illustrated in Figure 2.24.

The severe damage to the wing of a jet airliner Boeing 737-800 due to bird strike is illustrated in Figure 2.25 [22].

Figure 2.26 illustrates how severely birds can damage the wings of small aircraft.

Other examples of damage to small aircraft are given in Figures 2.27–2.29.

Figure 2.27 illustrates the damage to a Piper PA-28 Cherokee after striking a Canada goose during its final approach (50 ft AGL) on 21 April 2014. The aircraft experienced substantial damage and was out of service for 720 hours for repair [15].

Figure 2.28 illustrates the damage to a nearly brand new Cessna aircraft (only 80 flight hours) when it struck a large bird.

Figure 2.29 illustrates the damage to a Diamond 20 flying at 120 knots when it struck a tundra swan at 2000 ft AGL [15]. Costs of repair exceeded US$ 32 000.

Figure 2.30 shows the dent with a covering rupture (120 mm) on the right half-wing of an Antonov An-24 aircraft because of a strike with a mallard drake [18].

2.5.6 Empennage

The empennage includes both horizontal and vertical tails. Again, the FAA (only) added an extra requirement (14 CFR Part 25-631) that an aircraft must be capable of a safe flight and a subsequent normal landing after the empennage structure has been impacted by a 3.6 kg bird at cruise speed (V_c) at mean sea level [23, 24].

Figure 2.24 Typical bird strike incidents with wings. *Source:* courtesy USDA-APHIS [15].

Figure 2.25 Damage to wing of Boeing 737-800 due to bird strike. *Source:* courtesy USDA-APHIS [14].

Figure 2.31 illustrates the damage to different empennage elements caused by bird impacts.

2.5.7 Power Plant

Bird strike with an aircraft's power plant is the most dangerous type of bird strike event. Numerous accidents with fatalities have been caused by engine failure due to a bird strike. Large modern jet engines are more vulnerable to bird strike than earlier smaller

Figure 2.26 Damage to wings of small aircraft due to bird strike. *Source:* courtesy USDA-APHIS [14].

Figure 2.27 Damage to wing of Piper PA-28. *Source:* courtesy USDA-APHIS [15].

piston engines. Moreover, most current aircraft are powered by two turbofan engines rather than three or four ones, as in the 1970s and 1980 [25, 26]. Though aircraft can safely land with only one engine, if the second engine also experiences a malfunction then flight safety becomes more critical.

Figure 2.28 Damage to wing of Cessna aircraft due to bird strike. *Source:* courtesy USDA-APHIS [15].

Figure 2.29 Damage to wing of Diamond 20. *Source:* courtesy USDA-APHIS [15].

Figure 2.30 Bird strike on An-24 aircraft. *Source:* courtesy Aviation Ornithology Group, Moscow, Russia [18].

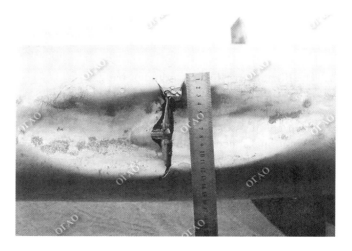

Figure 2.31 Damage to empennage caused by bird impacts. *Source:* courtesy USDA-APHIS [3].

If the bird enters the engine core, then an odor of burnt flesh may be smelt in the flight deck or passenger cabin from the bleed air.

If birds are ingested into engines, one or more of the following symptoms will be noticed:

1. The flight crew will hear a thud or bang.
2. The exhaust gas temperature (EGT) may exceed its typical value.
3. There will be a reduction or fluctuation in primary power parameters [25] such as:
   - fan/compressor (s) pressure ratio
   - rotational speeds of fan (N_1), compressor (N_2).
4. Abnormal fuel flow.
5. Abnormal engine vibration.
6. An engine stall (also called engine surge), which is, in fact, compressor surge. There are three possible engine responses:

- surge once and recover;
- surge once and continue until the flight crew take action;
- surge once and never recover, resulting in loss of power of that engine.

7. A flameout, as shown in Figure 2.32.
8. Fracture or disintegration of one or more fan blades. In such a case, the engine is likely to surge once and not recover.

In rare cases, bird ingestion of medium or large birds can be encountered by more than one engine. In such cases, multiple engine damage is likely and quick actions to stabilize the engines becomes a higher priority than if one engine is only involved.

Based on the FAR 33's requirements, during takeoff an engine must be able to withstand the strike of a 3.65 kg bird without catching fire, releasing hazardous fragments through the engine casing, or losing the ability to be shut down. It must also be able to withstand the simultaneous ingestion of several (up to four) smaller birds (around 1 kg) without losing more than 25% of thrust [27].

Here are details of some accidents where birds were ingested into an engine and flight crew landed safely without any fatalities.

On 3 June 1995 a Concorde aircraft (Figure 2.33) was in its landing phase (10 ft AGL) at John F. Kennedy Airport. One or two Canada geese were ingested into engines, which resulted in damage to engines #3 and #4 and cut several hydraulic lines and control cables. The cost of this bird strike was over US$ 7 million for repairs as well as US$ 5.3 million for the Port Authorities of New York and New Jersey as the runway was closed for several hours.

Figure 2.34 illustrates typical engine damage arising from bird strike for several engines.

Figure 2.35 illustrates the engine of an Airbus A320 that ingested a great blue heron during takeoff in 2002. Some fan blades and the nose cowl were damaged. The pilot made an emergency landing with the engine out. The runway was closed for some 38 minutes while fire trucks washed the debris from the runway [3].

Figure 2.32 Compressor surge and flame out [46]. *Source:* Sukhoi Su-57 prototype suffering a compressor stall at MAKS 2011. By Rulexip – Licensed Under CC BY-SA 3.0.

Figure 2.33 Concorde Aircraft during landing [45] *Source:* Concorde landing at Farnborough in September 1974 BY Steve Fitzgerald – Licensed Under: GFDL 1.2.

Figure 2.34 Typical engine damage arising from bird strike for several engines. *Source:* courtesy USDA-APHIS [14].

Figure 2.35 Damage to fan blades of an engine installed on Airbus A320. *Source:* Courtesy USDA-APHIS [3].

Figure 2.36 illustrates the damage experienced by a turbofan engine powering a Boeing B-767 aircraft as one or more double-crested cormorants were ingested into the engine #2 (also in 2002). Parts of the engine were broken and penetrated the engine casing, resulting in an uncontained engine failure [2].

Figure 2.37 illustrates the damage to fan blades of turbofan engine #2 of Fokker-100 aircraft arising from Canada geese ingestion in September 2004 [3].

Figure 2.38 illustrates damage to turbofan engine #1 of an MD-80 aircraft that ingested at least one double-crested cormorant on 16 September 2004 [3].

Figure 2.39 illustrates the damage to fan blades of #1 engine of an Airbus A330 aircraft due to a strike with a red-tailed hawk in March 2015. Four fan blades and seven exit guide vanes were replaced. The repair costs were US$ 1.5 million.

Figure 2.36 Blade damage to engine #2 of Boeing-767. *Source:* courtesy USDA-APHIS [3].

Figure 2.37 Blade damage to engine #2 of Fokker-100. *Source:* courtesy USDA-APHIS [3].

Figure 2.38 Damage to engine #1 of MD-80 aircraft in 2004. *Source:* courtesy USDA-APHIS [3].

Figure 2.39 Damaged fan blades of turbofan engine #1 powering Airbus A330. *Source:* courtesy USDA-APHIS [15].

Figure 2.40 illustrates the damage to a turbofan engine that was destroyed by a burrowing owl [29].

Figure 2.41 illustrates the damage experienced by the turbofan engine powering a Hawker 800 business jet that struck a flock of double-crested cormorants during takeoff

Figure 2.40 Damage to engine due to a flock of burrowing owls. *Source:* courtesy NTSB [28].

Figure 2.41 Damage to engine powering Hawker 800 business jet struck by a flock of birds. *Source:* courtesy USDA-APHIS [15].

(700 ft AGL) in April 2014. The multiple impacts of these 5 lb birds damaged engine #1 as well as the nose, fuselage, and landing gear. The pilot managed to land safely. The aircraft was out of service for 168 hours and the repair costs were at least US$ 825 000.

Figure 2.42 illustrates the damaged blades of the turbofan engine installed in a Boeing 767 that was struck by several Canada geese (400 ft AGL) in June 2013. The engine #2 exhibited severe vibrations, and the wing flaps were also damaged. The pilot made an emergency landing. Parts of two Canada geese and two fan blades were recovered from the runway. Repair costs were US$ 2.9 million.

Figure 2.43 illustrates the damage to the 1st and 2nd stage blades of a low-pressure compressor engine AI-25 of a Yak-40 aircraft after a strike with a middle-sized bird of prey [18]. The result of the strike and rise in the out-coming air temperature was that

Figure 2.42 Damage to engine of Boeing 767 in 2013. *Source:* courtesy USDA-APHIS [15].

 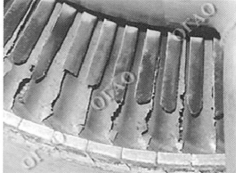

Figure 2.43 Damage to 1st and 2nd stage blades of a low-pressure compressor engine AI-25 of a Yak-40. *Source:* courtesy Aviation Ornithology Group, Moscow, Russia [18].

the turbine blades broke and the engine failed during the flight. This led to a forced emergency landing and a preschedule removal of the engine.

Figure 2.44 illustrates the $140 \times 30 \times 10$ mm dent on the AI-24 engine's air inlet on an An-24 aircraft as a result of a strike with a gull during a takeoff run at a speed of $180–200$ km h^{-1}.

Perhaps we should conclude this section with what is called the "Miracle on the Hudson." On 15 January 2009 an Airbus A320-214 (US Airways Flight #1549) was departing La Guardia airport at 15:30 Eastern Time and heading to Charlotte Douglas International Airport (CLT), North Carolina. During the initial climb, both engines were operating normally, but suddenly the aircraft encountered a flock of Canada geese. One Canada goose was ingested into each engine. The engines suffered from mechanical damage and the aircraft lost all engine thrust [29]. Unable to reach any airport, pilots Chesley Sullenberger and Jeffrey Skiles glided the plane to a water landing in the Hudson River off midtown Manhattan. The flight track is illustrated in Figure 2.45.

Figure 2.44 The 140×30×10 mm dent on the AI-24 engine's air inlet of An-24 aircraft. *Source:* courtesy Aviation Ornithology Group, Moscow, Russia [18].

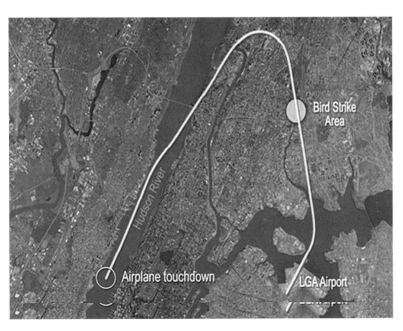

Figure 2.45 Flight track of the aircraft. *Source:* courtesy NTIS [29].

All 155 occupants (150 passengers and 5 crewmembers) were safely evacuated by nearby boats (Figures 2.46 and 2.47). No fatalities were recorded. However, five people were seriously injured (one cabin crew and four passengers) while 95 passengers had minor injuries. The remaining 55 persons (2 flight crew, 2 cabin crew, and 51 passengers) had no injuries.

The airliner was partially submerged and slowly sank (Figure 2.48).

Figure 2.46 Evacuating passengers from US Airways Flight 1549 using inflatable rafts. *Source:* courtesy NTIS [29].

Figure 2.47 Evacuating US Airways Flight 1549. *Source:* courtesy NTIS [29].

Figure 2.48 Sinking of US Airways Flight 1549. *Source:* courtesy NTIS [29].

The right engine remained attached to the wing and so was pulled from the river with the rest of the aircraft. However, the left engine was found to be separated from the wing. It splashed down into the Hudson River and was recovered eight days after the accident (Figure 2.49).

A single feather was recovered from the wing and sent to the Smithsonian Institution, which confirmed that the impacting bird was a Canada goose. Moreover, DNA testing on other bird remains also showed that Canada geese were the cause of the accident.

2.5.8 Propeller

On 10 August 2014, US Airways de Havilland Dash 8-100 aircraft (registration N815EX) flight number 4206, was climbing out of Harrisburg, PA, when the aircraft flew through a flock of geese and received a bird strike. It resulted in damage to the right-hand propeller, dent in the fuselage skin and a broken passenger window; Figure 2.50. The aircraft returned to Harrisburg for a safe landing [15, 32]. The damaged propeller caused strong engine vibrations and pilot aborted takeoff at 40 knots.

2.5.9 V-22 Osprey as a Military Example

The Bell Boeing V-22 Osprey; Figure 2.51 is an American multi-mission military aircraft. It is powered by two turboprop engines. However, the conventional propellers may be rotated through 90° for vertical takeoff and landing (VTOL). The propellers may then be tilted to develop an inclined thrust force for climbing. Finally, the propellers are rotated to have a horizontal axis for cruising.

In October 2005, during a low-level flight of Osprey aircraft over the Sierra Nevada mountains in California (200 ft AGL and about 220 knots), pilots saw a flash. The flash was due to a bird strike. Birds had stuck the hydraulic pump (Figure 2.52). The hydraulic pump that the bird is wedged behind produces 5000 psi and it is used to rotate the

Figure 2.49 Recovering of an engine from US Airways Flight 1549. *Source:* courtesy NTIS [29].

nacelles from aircraft mode to helicopter mode (Figure 2.53). Luckily, no hydraulic lines were ruptured and the damage was only cosmetic [30, 40].

2.5.10   Other Strikes to Aircraft Instruments

The following instruments are also susceptible to bird strike:

- the angle of attack AOA, used for stall warning;
- the total air temperature TAT probes, used for thrust setting;
- the pitot probes, used for airspeed measurement;
- antenna, glideslope, and localizer for runway guidance in meteorological conditions that require the use of instruments;
- WXR, weather radar, and wind shear instruments.

Figure 2.50 Damage to propeller blade of a DASH DHC8 aircraft. *Source:* Courtesy USDA-APHIS [16].

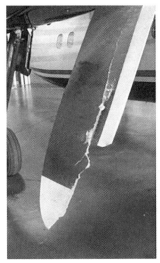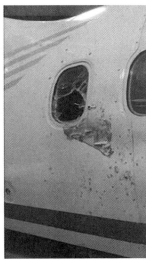

Figure 2.51 V-22 Osprey aircraft [40].

Figure 2.52 Bird strike with hydraulic pump of V-22 Osprey aircraft [40].

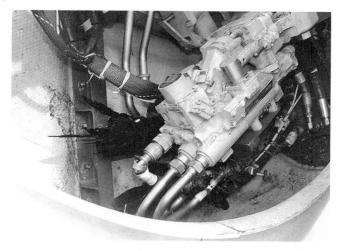

Figure 2.53 Bird wedged behind the hydraulic pump of V-22 Osprey aircraft [40].

2.6 Helicopters

The FAA Amendment 29-40 for certification AC 29.631, states that:

A. *For helicopters weighing more than 7000 lb, windshields must withstand a 2.2-lb (1 kg) at a relative velocity equal to the lesser of V_{NE} or V_H at altitudes up to 8000 ft without penetration. Two categories are identified*

- *For Category A certification, [31] the rotorcraft must be capable of continued safe flight and landing after the described bird strike.*
- *For Category B certification, the rotorcraft must be capable of a safe landing after the bird strike.*
- *Areas of the impact that are of particular interest include windshield, flight control surfaces (which includes main and tail rotors) and exposed flight control system components [33].*

B. But no bird strike safety standards exist for helicopters weighing less than 7000 lb. This represents about 90% of the U.S. helicopter fleet, including all tour and medical helicopters.

The European Aviation Safety Agency (EASA) explained the first reasons for these low certification requirements as:

- Low speed of helicopters as compared to fixed-wing aircraft
- High noise signature

Birds can avoid collisions and thus bird strikes certification was not identified as an immediate safety concern. However, recent helicopters fly faster and are less noisy. This helicopter features together with its more night operations, makes the probability of

Figure 2.54 Typical damage to helicopters as a result of bird strike. *Source:* courtesy USDA-APHIS [38].

helicopter bird strikes much higher and the need for bird strike certification of different sizes of the helicopter is mandatory.

EASA recommended that certification for small rotorcraft are necessary [21]. However, since changes in the regulations may take some time to be effective, the use of helmets and visors might, therefore, represent a more practical and timely option.

Figure 2.54 illustrates the typical damage to helicopters because of a bird strike.

On January 17, 2009, a Baptist Health MedFlight flying to Little Rock, it struck a flock of birds resulting in breaking the windshield. The pilot managed to make an emergency landing; Figure 2.55 [20].

A common crane bird (sometimes called Eurasian crane) collided with the windshield of a Sikorsky UH-60 Black Hawk helicopter [3]. Figures 2.56 and 2.57 show the damage to aircraft from both its outside and inside.

In April 2005, a Bell 407 air ambulance helicopter collided with three blue-winged teals during a flight (at 1000 ft AGL) in South Dakota, USA [34]. Figure 2.58 illustrates the shattered windshield, glass and duck blood splattered through the aircraft. Temporarily the pilot was blind, but he recovered soon and made an emergency landing on a nearby road.

Figure 2.59 illustrates a Mi-8 helicopter after striking a large bird. The bird strike destroyed the right full-metal frame of the heated window glass together with right non-heated glass and right heated glass with partial penetration into the cabin [18].

Figure 2.55 Damage to windshield of MedFlight helicopter. *Source:* courtesy USDA-APHIS [20].

Figure 2.56 Windshield of a Sikorsky UH-60 Black Hawk after impact with a common crane bird. *Source:* courtesy USDA-APHIS [3].

Figure 2.57 Inside the Sikorsky UH-60 Black Hawk after impact with a common crane bird. *Source:* courtesy USDA-APHIS [3].

Figure 2.58 Bird strike with a Bell 407 air ambulance helicopter. *Source:* courtesy USDA-APHIS [34].

2.7 Some Accident Data

Hereafter, some details of bird accidents for both fixed and rotary wings will be emphasized.

2.7.1 Fixed-Wing Aircraft

Table 2.1 lists the details of some bird strike accidents that resulted in structural damage only and no fatalities [9].

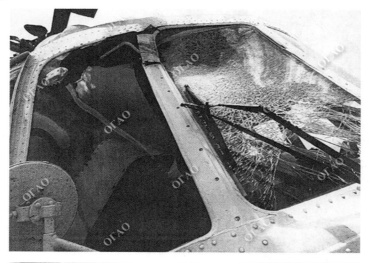

Figure 2.59 Bird strike with a Mi-8 helicopter. *Source:* courtesy Aviation Ornithology Group, Moscow, Russia [18].

2.7.2 Rotary-Wing Aircraft (Helicopters)

Table 2.2 lists the details of some bird strike accidents that resulted in structural damage only and no fatalities [9].

Table 2.1 Details of some bird strike accidents: fixed-wing aircraft [9].

Date	Location	Aircraft	Altitude (ft)	Indicated air speed (knots-KIAS)	Bird species	Mass (kg)	Impact KE (J)	Synopsis
15 August 1962	Lahore, Pakistan	Douglas DC3	8000 (cruise)	130	"Vulture"	5	15 046	Indian airlines flight between Kabul and Amritsar: a vulture penetrated the windshield
16 April 1972	Atlantic City, USA	Mitsubishi MU2		260	"geese"	1	8 945	Geese hit windshield
07 April 1981	Cincinnati, USA	Lear 23	3800	184	Common Loon	3.7	19 108	a) Damaged Copilot's windshield b) Engine 2 damaged by windshield debris c) Engine 2 shut down
06 August 1981	Musiars, Kenya	Cessna 402		140	Ruppel's Griffon Vulture	7.5	19 452	Vulture penetrated the windshield
15 July 1994	Florida, USA	Cessna 172	200	110	Pelican	7	11 298	A large bird collided with windshield causing windshield shattering. The pilot was injured in the face.
7 May 2018	Harare, Zimbabwe	Embraer ERJ-190	Still on the runway					Severe vibrations as the aircraft climbed out

Table 2.2 Details of some bird strike accidents: rotary-wing aircraft [9].

Date	Location	Aircraft	Altitude (ft)	Indicated air speed (knots-KIAS)	Bird species	Mass (kg)	Impact KE (J)	Synopsis
21 January 1985	Honolulu, USA	Hughes 369	400	130	A large flock of white birds			An extreme vibration developed due to a bird strike. The helicopter touched down on the water, rolled over & sank. The pilot escaped & swam to shore without injury.
30 May 1990	Louisiana, USA	Schweizer 269	800		A flock of birds (unidentified)			Main rotor blade flexed down and struck the tail boom. Aircraft became uncontrollable, thus struck the ground in a low nose attitude and rolled on its left side.
24 March 1993	USA	Bell 47	100	60	*Unknown bird*			Bird impacted the tail rotor which led to a loud bang and a vibration in the rudder pedals; then all yaw control was lost.

References

1 Knežević, J. (2014). Bird strike as a mechanism of the motion in Mirce mechanics. *Journal of Applied Engineering Science* 12 (3): 298.
2 Kazda, A. and Caves, E. (2007). *Airport Design and Operation*. Oxford: Elsevier.
3 Cleary, E.C. and Dolbeer, R.A. (2005). *Wildlife Hazard Management at Airports: A Manual for Airport Personnel*, 2e.
4 ACI World Operational Safety Subcommittee (2005). *Aerodrome Bird Hazard Prevention and Wildlife Management Handbook*. Geneva: ACI World Headquarters.
5 Gotoh, H., Takezawa, M., and Maeno, Y. (2012). Risk analysis of airplane accidents due to bird strikes using Monte Carlo simulations. *WIT Transactions on Information and Communication Technologies* 44: 393–403. https://doi.org/10.2495/RISK120331.
6 Aircraft Accident and Incident Investigation. Annex 13, International Civil Aviation Organization. http://www.emsa.europa.eu/retro/Docs/marine_casualties/annex_13.pdf (accessed 29 December 2018).
7 (1987). *Accident/Incident Reporting Manual (Adrep Manual)*, 2e. International Civil Aviation Organization (Doc 9156-AN/900) http://dgca.gov.in/intradgca/intra/icaodocs/Doc%209156%20-%20ADREP%20Manual%20Ed%202%20(En).pdf (accessed 29 December 2018).
8 Tiwari, A. (2015). What Really Happens When A Bird Hits An Airplane? https://www.scienceabc.com/pure-sciences/what-really-happens-when-a-bird-hits-strike-an-airplane.html (accessed 29 December 2018).
9 Bird Strike Damage & Windshield Bird Strike, ATKINS Final Report, 5078609-rep-03, Version 1.1. https://www.easa.europa.eu/system/files/dfu/Final%20report%20Bird%20Strike%20Study.pdf (accessed 29 December 2018).
10 Thorpe, J. (2003). Fatalities and destroyed civil aircraft due to bird strikes, 1912–2002. International Bird Strike Committee, IBSC 26, Warsaw. http://www.int-birdstrike.org/Warsaw_Papers/IBSC26%20WPSA1.pdf (accessed 29 December 2018).
11 Milson, T.P. and Horton, N. (1995). Birdstrike: An assessment of the hazard on UK civil aerodromes 1976–1990. Central Science Laboratory Report, Sand Hutton, York, UK.
12 Nave, C.R. (2016). HyperPhysics, Department of Physics and Astronomy, Georgia State University. http://hyperphysics.phy-astr.gsu.edu/hbase/impulse.html#c3 (accessed 29 December 2018).
13 Martin, H. (2013). Bird control problem and bird strike analysis of Czech and Slovak airports. *Transport Problems* 8 (3): 33–41.
14 Dolbeer, R.A., Weller, J.R., Anderson, A.L., and Begier, M.J. (2016). Wildlife Strikes to Civil Aircraft in the United States, 1990–2015, FAA report, November 2016. https://wildlife.faa.gov/downloads/Wildlife-Strike-Report-1990-2015.pdf (accessed 29 December 2018).
15 Dolbeer, R.A., Wright, S.E., Weller, J.R., Anderson, A.L., and Begier, M.J. (2015). Wildlife Strikes to Civil Aircraft in the United States, 1990–2014, Wildlife Strike Database, Serial Report Number 21. http://www.fwspubs.org/doi/suppl/10.3996/022017-JFWM-019/suppl_file/10.3996022017-jfwm-019.s8.pdf (accessed 29 December 2018).
16 Ahram Online (2016). EgyptAir passengers arrive in Cairo after bird strike delays flight from London, Sunday 13 March 2016. https://www.independent.co.uk/news/uk/home-news/bird-hits-passenger-jet-landing-at-heathrow-leaving-a-large-bloody-dent-on-its-nose-a6928781.html (accessed 14 January 2019).

17 Aviation News (2016). *Bird Strike Destroys Egyptian Plane's Nose.* https://www
.aviationcv.com/aviation-blog/2016/bird-strike-destroys-egyptian-planes-nose
(accessed 14 January 2019).

18 Aircraft Damages Caused by Birds, Aviation Ornithology Group, Moscow. http://www
.otpugivanie.narod.ru/damage/eng.html (accessed 29 December 2018).

19 Cleary, E.C. and Dickey, A. (2010). *Guidebook for Addressing Aircraft/Wildlife Hazards
at General Aviation Airports.* Washington, DC: The National Academies Press Airport
Cooperative Research Program, ACRP Report 32, 2010. https://doi.org/10.17226/22949
(accessed 29 December 2018).

20 Nicholson, R. and Reed, W.S. (2011). Strategies for Prevention of Bird-Strike Events,
AERO QTR_03.11, Boeing. http://www.boeing.com/commercial/aeromagazine/
articles/2011_q3/4/ (accessed 29 December 2018).

21 Hutchinson, J. (2015). Mail Online www.dailymail.co.uk/travel/travel_news/
article-3070051/Incredible-photos-reveal-aircraft-s-nose-COLLAPSED-bird-flew-
Turkish-Airlines-plane-carrying-125-passengers.html#ixzz4SHvQHgjp (accessed
29 December 2018).

22 Masterson, S. (2016). Bird Strike Requirements for Transport Category Airplanes:
Compliance by Analysis. Presented to Analytical Methods in Aircraft Certification
Workshop, 10 August 2016. https://www.niar.wichita.edu/niarfaa/Portals/0/FAA%20Bird
%20strike%20-%20Cert%20by%20Analysis%20-%20Suzanne%20Masterson.pdf (accessed
29 December 2018).

23 Skybrary (2018). Aircraft Certification for Bird Strike Risk. http://www.skybrary.aero/
index.php/Aircraft_Certification_for_Bird_Strike_Risk (accessed 29 December 2018).

24 Ugrčić, M., Maksimović, S.M., Stamenković, D.P. et al. (2015). Finite element modeling
of wing bird strike. *FME Transactions* 43 (1): 76–81.

25 Elsayed, A.F. (2008). *Aircraft Propulsion, and Gas Turbine Engines.* Boca Raton, FL:
Taylor & Francis, CRC Press.

26 Elsayed, A.F. (2016). *Fundamentals of Aircraft and Rocket Propulsion.* London, UK:
Springer.

27 Airbus (2006). Flight Operations Briefing Notes: Supplementary Techniques, Handling
Engine Malfunctions. http://www.smartcockpit.com/docs/Handling_Engine_
Malfunctions.pdf (accessed 29 December 2018).

28 DeFusco, R.P. and Unangst, E.T. (2013). Airport Wildlife Population Management:
A Synthesis of Airport Practice. Airport Cooperative Research Program, Synthesis 39.
https://www.nap.edu/read/22599/chapter/1 (accessed 29 December 2018).

29 National Transportation Safety Board (2010). Loss of Thrust in Both Engines After
Encountering a Flock of Birds and Subsequent Ditching on the Hudson River US
Airways Flight 1549 Airbus A320-214, N106US Weehawken, New Jersey. https://www
.ntsb.gov/news/events/Pages/Loss_of_Thrust_in_Both_Engines_After_Encountering_
a_Flock_of_Birds_and_Subsequent_Ditching_on_the_Hudson_River_US_Airways.aspx
(accessed 27 December 2018).

30 Private communication with US Air Force Academy (USAFA), Springfield, CO (2013).

31 European Aviation Safety Agency (2012). Certification Specifications for Large
Rotorcraft CS-29. https://www.easa.europa.eu/document-library/certification-
specifications/cs-29-amendment-3 (accessed 29 December 2018).

32 The Aviation Herald (2017). www.avherald.com, Last Update: Monday, June 5th, 15:10Z

33 US Department of Transportation (2014). Certification Of Transport: Category Rotorcraft, Advisory Circular No. 29-2C. https://www.faa.gov/documentLibrary/ media/Advisory_Circular/AC_29-2C.pdf (accessed 29 December 2018).

34 Cleary, E.C., Dolbeer, R.A., and Wright, S.E. (2006). Wildlife Strikes to Civil Aircraft in the United States 1990–2005. http://digitalcommons.unl.edu/cgi/viewcontent.cgi? article=1006&context=birdstrikeother (accessed 29 December 2018).

35 Uhlfelde, E. (2013). *Bloody Skies: The Fight To Reduce Deadly Bird-Plane Collisions*, National Gepgraphic. https://news.nationalgeographic.com/news/2013/10/131108-aircraft-bird-strikes-faa-radar-science/ (accessed 14 January 2019).

36 Dolbeer, R.A., Wright, S.E., Weller, J. and Begier, M.J. (2013). Wildlife Strikes to Civil Aircraft in the United States, 1990–2012. Serial Report Number 19. FAA and USDA-APHIS.

37 National Transportation Safety Board (2017). Preventing Catastrophic Failure of Pratt & Whitney Canada JT15D-5 Engines Following Birdstrike or Foreign Object Ingestion. https://www.ntsb.gov/investigations/AccidentReports/Reports/ASR1703.pdf (accessed 29 December 2018).

38 Keirn, G. (2013). Helicopters and Bird Strikes; Results from First Analysis Available Online. https://www.usda.gov/media/blog/2013/06/06/helicopters-and-bird-strikes-results-first-analysis-available-online (accessed 29 December 2019).

39 Air Combat Command, C-130 and Hawk inside nosecone. https://commons .wikimedia.org/wiki/File:C-130_and_Hawk_inside_nosecone.jpg (accessed 29 December 2018).

40 Private communication, Lt Col Sean Londrigan, DFEM Deputy Dept. Head, US Air Force Academy.

41 Yadav, B.K. (2017). Aircraft collisions and bird strikes in Nepal between 1946–2016: a case study. *Journal of Aeronautics and Aerospace Engineering* 6 (4): 203. https://doi .org/10.4172/2168-9792.1000203 (accessed 29 December 2018).

42 CHALLENGER 604 – 5483 – BIRD STRIKE ACCIDENT 2008 APR 08 https://avtales .wordpress.com/2008/04/08/challenger-604-5483-bird-strike-accident-2008-apr-08 (accessed 29 December 2018).

43 Full Size Photos from the Show Low Bird Strike Now Available (2009). http://www .birdstrikenews.com/2009/12/full-size-photos-from-show-low-bird.html (accessed 29 December 2018).

44 Air Combat Command (2005). F-16 canopy after a bird strike. https://en.wikipedia.org/ wiki/Bird_strike#/media/File:F16_after_bird_strike.jpg (accessed 29 December 2018).

45 https://upload.wikimedia.org/wikipedia/commons/1/11/Concorde_landing_ Farnborough_Fitzgerald.jpg (accessed 29 December 2018).

46 Rulexip (2011). Sukhoi Su-57 prototype suffering a compressor stall at MAKS 2011. https://en.wikipedia.org/wiki/Compressor_stall#/media/File:PAK_FA_T_50_ compressor_stall_on_MAKS_2011.jpg (accessed 29 December 2018).

47 Brotak, Ed. (2018). FAA: 665 helicopter bird strikes over the last 3 years. Vertical Magazine. https://www.verticalmag.com/news/faa-665-helicopter-bird-strikes-over-last-3-years/ (accessed 14 January 2019).

3

Statistics for Different Aspects of Bird Strikes

Bird Strike in Aviation: Statistics, Analysis and Management, First Edition. Ahmed F. El-Sayed.
© 2019 John Wiley & Sons Ltd. Published 2019 by John Wiley & Sons Ltd.

3.1   Introduction

Throughout history, humans have been inspired by the beauty of birds and their ability to fly. Bird flight dates to about 150 million years ago, while humans first began to share the air with birds only in 1903. Unfortunately, when aircraft and birds share the sky then collisions occur [1].

The number of birds in the world is approximated as 100 000 000 000 (1×10^{11}), while the number of flying aircraft has a peak of 10 000. Consequently, the likelihood of bird strike is high.

Though bird strike is an issue as old as aviation, its hazard has not been diminished. This chapter will provide the reader with many statistics about bird strikes.

Bird strike statistics are collected by civil aviation authorities/agencies in many countries. Some countries maintain separate military and civil strike databases, while others maintain combined databases.

Three major wildlife strike databases are available:

- United States Federal Aviation Administration (FAA) database;
- Transport Canada bird/mammal strike database;
- International Civil Aviation Organization (ICAO) database.

Apart from the above-stated databases, two well-known researchers in the bird strike field have independently developed separate wildlife strike accident databases.

- John Thorpe: retired from the Civil Aviation Authority in the UK, former Chairman of the Bird Strike Committee Europe, and honorary Chairman of International Bird Strike Committee (IBSC). He has compiled a worldwide database of all known serious civilian aircraft accidents involving birds.
- Dr W. John Richardson of LGL Limited, Canada. He has created a database of military aircraft incidents involving birds.

3.2   Statistics for Bird Strike

Collecting and reporting reliable data on collisions between birds and aircraft is the first crucial step in identifying factors that contribute to incidents/accidents of bird/aircraft collisions. Arranging such information in databases is the second step. Performing statistical analyses for bird strike data based on different parameters comes as a third step.

The FAA has a standard form: FAA 5200-7, Bird/Other Wildlife Strike Report. Pilots, airport operators, and aircraft maintenance personnel should report these strikes. Voluntary reporting of bird/other wildlife strikes with aircraft by anyone who knows of a strike is encouraged. It is important to include as much information as possible on the FAA Form 5200-7.

To improve the ease of reporting, strikes can also be reported via the Internet (http://wildlife-mitigation.tc.faa.gov).

The FAA Bird/Wildlife Strike Database is managed by the Wildlife Services program of the US Department of Agriculture (USDA). All strike reports are reviewed by the staff wildlife biologists at the FAA and, finally, a database manager edits all the strike reports for the same strike before entering the data.

Reporting of bird strike incidents is not uniform across all continents. The best reporting is available in the UK, the USA, and Canada. It is estimated, based on previous FERA (UK Food & Environment Research Agency) Bird Management Unit experience [2], that the UK and North America data obtained is nearly 50% of all worldwide bird strike reports. Moreover, regions such as South America, Russia, and China do not routinely contribute data to the ICAO database.

Some brief statistics concerning wildlife strike with transport aircraft up to 2016 may be outlined as follows:

- 169 856 wildlife strikes;
- 97% involved birds;
- 63% during the day;
- 52% of strikes were between July and October;
- 61% were during the arrival phases of a flight.

Concerning general aviation (GA) aircraft, the following statistics are also identified:

- 73% occurred below 500 ft AGL;
- 88% took place within the airport environment (below 1500 ft AGL);
- 97% took place below 3500 ft AGL (the bird-rich zone);
- the rate of damaging strikes has not declined since 2000;
- the rate of damaging strikes has increased outside the airport environment.

Statistics collected include the following parameters:

- number of birds colliding with aircraft;
- types of aircraft most likely to be struck;
- the parts of the aircraft most likely to be struck;
- the phase of flight when strikes are most likely to occur;
- the altitude at which strikes occur;
- problematic times of the day, month, and year;
- yearly strike trends by location;

- dangerous wildlife species;
- effects of strikes on aircraft;
- percentage of strikes that are damaging and affect the flight;
- costs associated with strikes.

3.3 Classifying Bird Strikes

Bird strike reports can involve single or multiple strikes. Three categories are used in bird strike reporting, namely:

- single or multiple large bird(s);
- small numbers of medium-sized birds (between 2 and 10 birds);
- large flocks of relatively small birds (more than 10 birds).

3.3.1 Single or Multiple Large Bird(s)

A single large bird may threaten an aircraft if ingested into the engine or impacting the radome of the aircraft.

3.3.2 Relatively Small Numbers of Medium-Sized Birds (2–10 Birds)

Though these involve few birds, if they are ingested into the engine they may cause a compressor surge and damage different rotating parts. An example of such a case is a Thomson Airways Boeing 757 during takeoff from Manchester en route to Lanzarote at 9.15 a.m. on 29 April 2007. Two herons were sucked into the starboard engine; the engine caught fire, forcing the pilot to make an emergency landing.

3.3.3 Large Flocks of Relatively Small Birds (Greater Than 10 Birds)

Generally, small birds need to travel in flocks to avoid predators. However, such flocks may cause catastrophic accidents. Figure 3.1 illustrates a flock of blackbirds threatening a landing aircraft.

The number of bird strike reports worldwide in the interval from 1990 to 2007 obtained from different sources in the UK (Civil Aviation Authority), Canada (Transport Canada), and the USA (US Department of Agriculture), is 94 743 [2]. The number of these that caused some level of aircraft damage is only 10 750. These reports involved the airframe, where both aircraft type and bird species are known. The number of collisions by single and multiple bird species that resulted in some level of damage to the aircraft is given in Table 3.1.

3.4 Classification of Birds Based on Critical Sites in the Aerodrome

Birds are classified based on the critical sites in the aerodrome they occupy and their movements into the path of flying aircraft in order to cause a strike and thus become hazardous. Six groups can be identified [4].

Figure 3.1 Flock of blackbirds. *Source:* courtesy National Academies Press [3].

Table 3.1 Single and multiple strikes [2].

Number of birds struck	1	2–10	>10	Total
Number of strikes	7704	2726	320	10750

3.4.1 Birds Flying or Soaring Over the Aerodrome or Approach Paths (100–4000 ft AGL)

Examples of birds flying or soaring over the aerodrome or approach paths are vultures (Figure 3.2), harriers, and kites [4]. These birds are hazardous to aircraft at the climb, cruise, descent, and approach flight phases;.

Figure 3.2 Vulture soaring at 100–4000 ft or above. *Source:* courtesy IBSC [4].

Figure 3.3 Northern Harrier hovering at 2200 ft. *Source:* courtesy IBSC [4].

Partridges   Stone Curlews

Figure 3.4 Birds squatting on the runway to rest. *Source:* courtesy IBSC [4].

3.4.2 Birds Flying, Sailing Low, or Hovering Over Active Runway and Shoulders (2200 ft AGL)

Examples of birds flying, sailing low, or hovering are harriers (Figure 3.3), kestrel, kites, and black-winged kites [4]. These birds are hazardous at the following flight phases: takeoff run, initial climb, final approach, and landing roll.

3.4.3 Birds Perching and Walking on Runway/Shoulders

Examples of birds perching and walking on runway/shoulders are partridges, stone curlews (Figure 3.4), harriers, kites, and lapwings [4]. These birds are hazardous to aircraft during landing, takeoff, and taxiing. If they are sucked into the engine, then serious engine damage may occur.

3.4.4 Birds Squatting on the Runway to Rest

An example of a bird that squats on the runway is the pariah kite (Figure 3.5). This group is like the previous one, and is hazardous to aircraft during landing, takeoff, and taxiing. They can also be sucked into the engine, causing damage.

Figure 3.5 Pariah Kites. *Source:* courtesy IBSC [4].

Figure 3.6 American Crow. *Source:* courtesy IBSC [4].

3.4.5 Birds Feeding on Live or Dead Insects or Animals on the Runway

Examples of birds that feed on the runway are kites, crows (Figure 3.6), and harriers. These birds are hazardous to aircraft during landing and takeoff flight phases.

3.4.6 Birds Perched on Runway Lights, Floodlight Towers, Electric Poles, and Other Perches

Examples of birds that perch on aerodrome equipment are kites, rollers, kestrels (Figure 3.7) and the black-winged kite [4].

Figure 3.7 Kestrel. *Source:* courtesy IBSC [4].

3.5 Bird Impact Resistance Regulation for Fixed-Wing Aircraft

The regulations that cover impact resistance are listed for different sizes of aircraft [2, 5, 6] and are discussed below.

3.5.1 Transport Aircraft (Airliners, Civilian, and Military Cargo)

These aircraft are those for which the maximum takeoff weight (MTOW) is greater than 5670 kg (12 500 lb).

3.5.1.1 Airframe

Both the FAA code (Pt 25) and EASA code (CS-25) agree on the following require-ments: all airframe modules, except the empennage, must be able to withstand without hazard (penetration or critical fragmentation) impact with a bird of 1.8 kg (4 lb) at V_c at sea level or 0.85 V_c at 8000 ft, whichever is the most critical [2].

The most significant difference between the US and European requirements is concerned with the empennage. The FAA have issued an additional requirement that empennage structure should withstand the impact of a 3.6 kg (8 lb) bird at cruise speed (V_C), while the EASA have no separate requirement for the empennage and keep the same structural requirement of 1.8 kg (4 lb) at V_C (CS-25).

3.5.1.2 Engines

Newly certified engines have to meet a range of EASA European engine ingestion tests (detailed in CS-E800, based on thrust generation) as follows:

- *No hazardous engine effect after impact with a single large bird of weight between 1.85 and 3.65 kg.*
- *For a large flocking bird, no more than 50% loss of thrust following ingestion of large mixed birds of between 1.85 and 2.5 kg and for 20 minutes must be capa-ble of thrust variation.*

- *For medium flocking birds, the engine must be capable of ingesting some birds of varying mass, without a reduction of the thrust of less than 75%.*
- *For small birds of weight 0.85 kg, ingestion of some birds must not result in loss of more than 25% thrust.*

Most recent airplanes meet these criteria, but previously produced engines cannot satisfy these new design requirements.

3.5.2 General Aviation Aircraft

General aviation aircraft include both lightweight executive jets of less than 5700 kg (12 500 lb) and planes in the "commuter" category, which are propeller-driven twin-engine airplanes having MTOW of 8618 kg (19 000 lb) or less and that have 19 or fewer seats. This group of aircraft will have been certificated to US Part 23 and European (CS-23) requirements for general aviation aircraft, which has no specific bird strike certification requirements (apart from windshields on commuter aircraft).

3.5.3 Light Non-Commuter Aircraft

Examples of light non-commuter aircraft are Cesena 206, Socata, and Extra EA-400/500, all of which that a capacity of five passengers. These planes have no bird strike-related certification requirements.

3.6 Bird Impact Resistance Regulation for Rotorcrafts

3.6.1 Large Rotorcraft

The "large rotor craft" category covers Cat A, with a weight greater than 9072 kg (20 000 lb) and 10 or more passenger seats, and Cat B, other weights. Based on EASA CS-29 and the FAA Federal Aviation Regulation (FAR) Part 29, the requirements for this type of aircraft are identical [2]. Both require that the aircraft can continue safe flight and landing or safe landing following impact with a single 1 kg bird at the greater of the maximum safe airspeed (V_{NE}) or maximum level-flight airspeed at rated power (V_H) (at up to 8000 ft).

3.6.2 Small Rotorcraft

For small rotorcraft (weight 3175 kg/7000 lb or less), neither the FAA Pt-27 nor the EASA CS-27 codes contain any requirement for protection against bird strike [2].

3.7 Statistics for Fixed-Wing Civilian Aircraft

Civilian fixed-wing aircraft are categorized as "Commercial Transport and General Aviation" aircraft. Both types are further divided into two groups based on their power plant; these are either turbine-powered or shaft-powered aircraft [7]. Turbine-powered aircraft are powered by either turbojet or turbofan engines, while shaft-powered aircraft are powered by either piston engines (normally small aircraft) or turboprop engines.

The parts of civil aircraft mostly impacted by birds are the nose, radome, windshield, engines, fuselage, wing, empennage, landing gear, instruments, and lights.

In a study by the ICAO on wildlife strikes with aircraft in 105 states and territories during the period 2008–2015, a total of 97 751 cases were reported [8]. The effect of the wildlife strike on the flight was reported 12 227 times. Out of those, 2501 had a clear indication of an effect on the flight, while the rest had no effect or no clear indication.

These 2501 cases represent 20% of the total number in which the effects were reported. These 2501 cases may be categorized as follows:

- precautionary landings, 1230 cases (49%);
- aborted takeoffs, 513 cases (21%);
- engine(s) shut down, 63 cases (3%);
- delayed flights, 211 cases (8.4%);
- declaring technical emergency, 54 cases (2.2%);
- aircraft forced to return, 137 cases (5.4%);
- unknown reasons, 305 cases (11%).

3.7.1 Critical Parts of Turbofan/Turbojet Aircraft

Figures 3.8–3.11 display the findings of different airliners manufactures or authorities regarding the percentages of bird strikes with different parts of an aircraft.

Figure 3.8 illustrates the findings of Airbus Industries [9].

Figure 3.9 illustrates the data of bird strikes published by the EASA [10].

The findings of the Boeing Company are listed by Nicholson and Reed [11].

According to the data registered for civil aviation aircraft belonging to Russian air companies in the interval 2002–2005 [12], the percentage of strikes with different parts of the plane is shown in Figure 3.10.

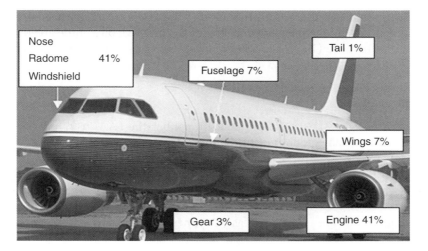

Figure 3.8 Percentages of bird strike, Airbus Industries.

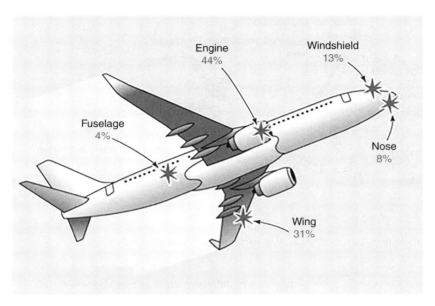

Figure 3.9 Percentages of bird strike, EASA.

Figure 3.10 Percentages of bird strike, Russian statistics. *Source:* courtesy Aviation Ornithology Group, Moscow, Russia [12].

The statistics for bird strikes recorded in the Asia-Pacific regions from 2006 to 2010 [1] is plotted in Figure 3.11.

A summary of the statistics of different companies/authorities displayed in Figures 3.8–3.11 is given in Table 3.2.

From Table 3.2, the following two parts of aircraft are shown to be the most critical:

- nose, radome, and windshield, 11.6–60%;
- engines, 16.8–48.9%.

From Table 3.2, it is interesting to state that both Boeing Company (representing the US airplane manufacturer) and the European authority EASA (European Aviation Safety Agency) report the same figures for bird strikes. The US and European results are illustrated in Figure 3.12.

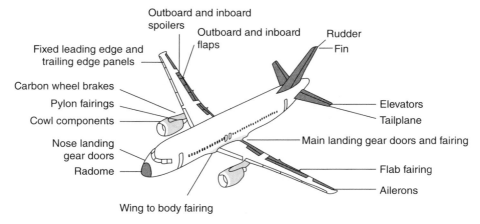

Figure 3.11 Parts of aircraft struck by birds, Asia- Pacific regions from 2006 to 2010.

Table 3.2 Summary of percentages of bird strikes with different parts of turbine-powered aircraft.

	Company/Authority					
Affected area	**Airbus (%)**	**EASA (%)**	**Boeing (%)**	**Russian companies (%)**	**Asia/Pacific region (%)**	**Average (%)**
Nose		8	8		14	
Radome	41			4.9	23	11.6–60
Windshield		13	13	6.7	23	
Engine 1					4.8	
Engine 2	41	44	44	48.9	2.4	16.8–48.9
Engine 3					9.6	
Fuselage	7	4	4	9.0	16	4–16
Landing gear	3			6.7	4.8	3–6.7
Wing	7	31	31	21.1		7–31
Tail	1			2.7		1–2.7
Lights					2.4	2.4

A detailed record for bird strikes with civil aircraft components from the USA (including US registered aircraft in foreign countries) over 26 years (1990–2015) is listed in Table 3.3 [13].

A recent study was performed to examine the effects of engine lighting color on bird strikes [14]. Two types of engine installations were examined – namely wing installation (identified by the authors in [14] as under-the-wing (alternatively identified by aero engine manufacturers as pod installation [7]) and fuselage installation (Figure 3.13). The Wildlife Strike Database of the US FAA was used to compare bird strikes reported to engine #1 (left-hand side, red lighting) to engine #2 (right-hand side, green lighting).

Aircraft location of bird strike damage in accidents (1999–2008)

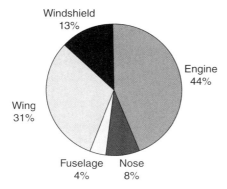

Figure 3.12 Location on the aircraft which was struck and damaged by bird(s), worldwide (1999–2008).

Table 3.3 US Civil aircraft components reported as being struck and damaged by birds in 26 years (1990–2015).

Aircraft component	Number struck	Total (%)	Number damaged	Total (%)
Windshield	23 503	16	1011	6
Nose	20 594	14	1037	6
Wing/rotor	19 845	14	3862	24
Radome	17 832	12	1550	10
Engine(s)	17 494	12	4516	28
Fuselage	17 210	12	676	4
Other	15 820	11	1302	8
Landing gear	6409	4	530	3
Propeller	3131	2	275	2
Tail	1861	1	659	4
Light	970	1	692	4
Total	14 466	100	16 110	100

In both engine installation positions (wing pod and fuselage), bird strikes with engine #1 were more common than those with engine #2 during day, night, and dawn/dusk flights.

Table 3.4 lists the number of bird strikes for turbofan engines installed on wings of some civil transport aircraft in the USA in the period January 1990–July 2015 [14]. There were 56 142 strike reports involving birds for the aircraft listed in Table 3.4. Data is also available for Boeing B767, B777, and B787 as well as Airbus A300/310. The data reported 49 178 strikes to aircraft parts other than engines and 321 had strikes to both engines. The remaining 6643 reports with a strike to engine #1 or #2 were used in the analyses.

(a)   (b)

Figure 3.13 Wing and fuselage installations. (a) Engines on wings, Boeing 737 [36]. (b) Engines on fuselage, Embraer ERJ-145 [37].

Table 3.4 Bird strike with turbofan engines installed to wings of some aircraft in the USA in the period January 1990–July 2015 [14].

Aircraft series	Day Engine #1	Day Engine #2	Night Engine #1	Night Engine #2	Dawn/Dusk Engine #1	Dawn/Dusk Engine #2	All times Engine #1	All times Engine #2
Boeing B-737	1212	1017	427	374	16	15	2124	1817
Airbus A-318 to 330	242	257	104	117	36	31	546	555
Boeing B-757	143	114	68	57	12	6	293	247
Embraer EMB-170/190	72	64	22	21	—	—	146	135
Dornier 328 J	6	5	—	—	1	—	7	5

Table 3.4 shows that aircraft subjected to most bird strikes is the Boeing B737. The reason for this is that the engines are very close to the ground. Other research looking at foreign object damage (FOD) confirms the same finding that the B737 is the most vulnerable aircraft to other FOD sources, as outlined in Chapter 1.

Table 3.5 lists the number of bird strikes for turbofan engines installed on the fuselage of some civil transport aircraft in the USA in the period January 1990–July 2015 [14]. There were 33 119 strike reports involving bird strikes with the aircraft listed above together with others of the following groups: Dassault DA-10/200, Hawker 800/4000, Beechcraft BE-400 BJET, Canadair/Bombardier CL-600/604, Fokker 100/F28, Gulfstream III/IV, Boeing B 717-200, Gulfstream 200/V, British Aerospace BAe-125-700/800, Mitsubishi MU-300, and miscellaneous. Of these, 30 822 had strikes to aircraft parts other than engines and 148 had strikes to both engines. The remaining 2149 reports with a strike to either engine #1 or #2 were used in the analyses.

The analysis summarized in Tables 3.4 and 3.5 provided evidence of a trend toward birds striking the left-hand side of aircraft, where a red navigation light is located, compared to the right-hand side, where a green light is located. Surprisingly, the trend was

Table 3.5 Bird strike with turbofan engines installed to the fuselage of some aircraft in the USA in the period January 1990–July 2015 [14].

Aircraft series	Day		Night		Dawn / Dusk		All times	
	Engine #1	Engine #2	Engine #1	Engine #2	Engine #1	Engine #2	Engine #1	Engine #2
McDonnell Douglas DC-9/MD-80s	155	88	81	50	29	26	294	193
Cessna citation	122	89	25	31	25	11	199	156
CRJ 100–900	97	100	40	34	11	13	164	165
Embraer EMB-135/145	78	69	28	25	14	10	144	124
Learjet-31/60	85	53	16	9	20	18	124	85

greatest during the day and not during the night and dusk/dawn, though these periods would be when aircraft lighting has the largest influence.

These findings suggest it would be sensible to modify red navigation lights to include shorter wavelengths, with the use of supplemental lights specifically designed for avian vision to enhance better detection and reduce bird strikes.

Finally, experts from different agencies in the USA, including the FAA and the US Department of Agriculture, and in Europe, including the EASA, expect that the risk, frequency, and severity of wildlife/aircraft collisions to grow over the next decade. This conclusion is based on increasing air traffic, increasing bird populations [3], and the adoption of twin-engine aircraft [9].

3.7.2 Critical Modules of Turboprop/Piston Aircraft

Small aircraft are powered by piston engines. These engines are then fitted with propellers to generate the necessary thrust force. Examples of piston-powered aircraft are the Cesena Skylane and Cesena Skyhawk. Many medium- and large-sized aircraft are powered by turboprop engines, which are also fitted with medium/large-sized propellers. The main difference between piston engines and turboprops engine is that the propeller installed to the turboprop engine generates most of the thrust force (80–90%) necessary for aircraft propulsion, while the propeller fitted to a piston engine generates the whole thrust force [7]. An example of a turboprop-powered aircraft is the Airbus A400 (Figure 3.14). In all cases, the propeller, together with other aircraft parts, may be impacted by birds.

Piston-powered aircraft are noisy and relatively slow, and thus birds can usually avoid these aircraft. Moreover, even if strikes occur they will cause little or no damage.

Similarly, the propellers in turboprop-powered aircraft are noisy and operate at moderate rotational speeds. Thus they are less vulnerable to bird strike damage compared with the fans in turbofan engines.

Figure 3.15 illustrates the location of bird strike damage on aircraft powered by turboprop engines in accidents during 2010.

Figure 3.14 Airbus A400, powered by four turboprop engines. *Source:* courtesy Airbus Industries.

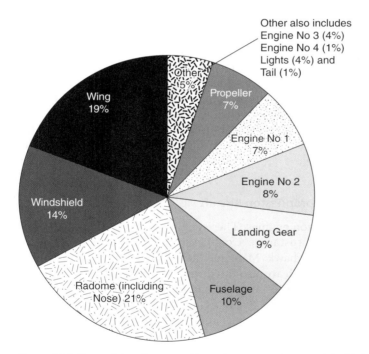

Figure 3.15 Parts of aircraft powered by turboprop engines struck by birds in 2010.

Table 3.6 illustrates a comparison between locations of bird strike damage from accidents for turbofan/turbojet-powered aircraft and those powered by turboprop engines. As outlined above, the noise generated by propellers in turboprop engines frightens birds and results in lower percentages of bird strike on the engines. Even the combined percentages of propeller and engine in turboprop cases (22–27%) are less than turbofan engines (44%).

Table 3.6 Damage to aircraft parts powered by turbofan/turbojet vs. turboprop engines.

Aircraft engine	Radome including nose (%)	Windshield (%)	Engine (%)	Wing (%)	Fuselage (%)	Propeller (%)	Landing gear (%)
Turbofan/ Turbojet	8	13	44	31	4	—	—
Turboprop	21	14	15–20	19	10	7	9

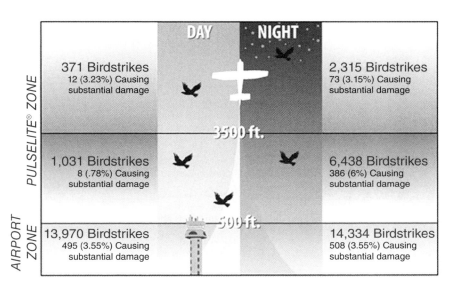

Figure 3.16 Survey of the number of strikes during the day or at night versus flight altitude. *Source:* courtesy USDA-APHIS [15].

3.7.3 Bird Strike Versus Altitude

Figure 3.16 illustrates the number of bird strikes during daylight and at night for flight altitudes up to 10 000 ft above ground level (AGL) from 1990 to 2002 [15]. The risk of bird strike increases with proximity to the ground. Moreover, bird strike during the night is much greater than during daylight.

As displayed in Figure 3.17, various sources state that nearly 95% of bird strikes occur at altitudes below 2500 ft AMSL (above mean sea level), while around 70% occur at altitudes below 200 ft [5]. Though various sources have different percentages for each altitude threshold, they all concur that most accidents take place very close to the ground.

Moreover, data collected from Directorate General of Civil Aviation of France (French DGAC) and identified by Airbus Industries [9] provides a detailed breakdown for bird strike close to the ground (Figure 3.18). It can be categorized as follows:

- 50–60% are encountered at altitudes less or equal to 50 ft;
- 30% for altitudes ranging from 50 to 500 ft;
- 10–20% occur at higher altitudes.

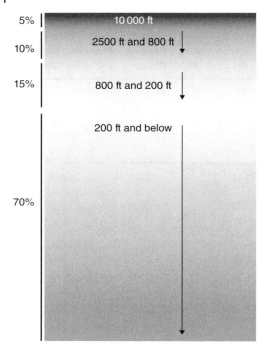

Figure 3.17 Percentages of bird strikes per different altitude bands above ground level.

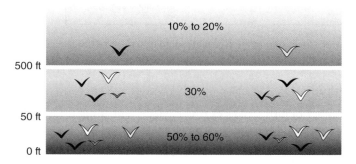

Figure 3.18 Bird strike versus altitude AGL. *Source:* courtesy Airbus [9].

However, the FAA wildlife hazard management manual for 2005 [16] categorized bird strike accidents against altitude as follows:

- less than 8% of strikes occur above 900 m (3000 ft);
- 61% occur at less than 30 m (100 ft);
- a few accidents occur at high altitudes, some as high as 6000 – 9000 m (20 000–30 000 ft).

Two astonishing cases for birds flying at very altitudes are as follows:

- bar-headed geese have been seen flying as high as 10 175 m (33,383 ft);
- a Rüppell's vulture flying at an altitude of 11 300 m (37,100 ft) collided with an aircraft over the Côte d'Ivoire.

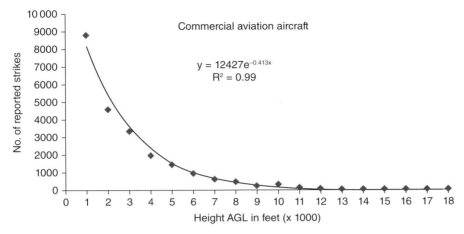

Figure 3.19 The number of reported bird strikes for commercial aviation aircraft versus altitude, USA, from 1990 to 2015. *Source:* Courtesy USDA-APHIS [13].

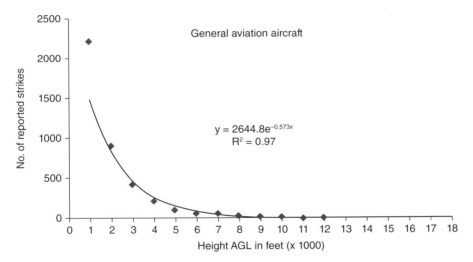

Figure 3.20 The number of reported bird strikes for general aviation aircraft versus altitude, USA, from 1990 to 2015. *Source:* courtesy USDA-APHIS [13].

Figures 3.19 and 3.20 illustrate the number of reported bird strikes for commercial aviation aircraft and general aviation aircraft, respectively, versus altitude at 1000-ft height intervals above ground level in the period from 1990 to 2015. A fitted empirical relationship for each case is written on each figure [13]. Both the empirical formulae show an exponential decay for the number of bird strikes with altitude.

Data from Figures 3.19 and 3.20 are given in Table 3.7, where the number of reported bird strikes to commercial and general aviation aircraft in the USA by height above ground level (AGL) in the period 1990–2015 is listed in 500 ft increments. There is a strong relationship between bird strike altitude and percentage of strikes causing damage. The greatest number is reported for bird strikes on commercial aircraft at sea level (33 100), which reduces to 296 for heights greater than 11 500 m. For general aviation

Table 3.7 Number of reported bird strikes to commercial and general aviation aircraft by height above ground level (AGL), in the USA, 1990–2015 [13].

Height of strike (AGL, in feet)	Commercial aviation aircraft		General aviation aircraft	
	All reported strikes	Strikes with damage	All reported strikes	Strikes with damage
0	33 100	1831	5537	644
1–500	24 611	1769	5471	1299
501–1500	8810	947	2217	981
1501–2500	4607	571	910	443
2501–3500	3361	375	417	205
3501–4500	1990	220	212	99
4501–5500	1470	175	101	46
5501–6500	961	122	62	31
6501–7500	663	84	52	20
7501–8500	490	73	29	17
8501–9500	264	34	18	10
9501–10 500	344	57	19	13
10 501–11 500	185	43	4	2
>11 500	296	90	26	17
Total	81 152	6391	15 075	3827

aircraft, the number of strikes at sea level is 5537 and reduces to 26 for heights greater than 11 500 m.

Figure 3.21 illustrates reported bird strikes for the US, Canada, and UK from 1990 to 2007 [2]. Two peaks can be seen; namely, at very low altitude (less than 100 ft) and

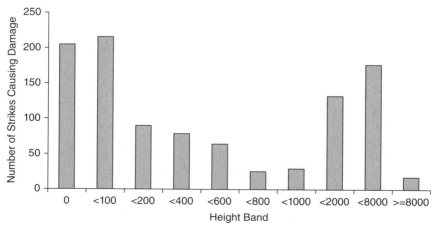

Figure 3.21 Number of damaging strikes at different height bands in USA, Canada, and the UK from 1990 to 2007. *Source:* courtesy USDA-APHIS [14].

higher altitudes (between 2000 and 8000 ft). The low altitude may be explained by a large number of birds, while the higher altitude peak is mainly due to the greater aircraft speed that results in a greater probability for damaging strikes.

3.7.4 Bird Strike by the Phase of Flight

Most bird strikes (90%, according to the ICAO) happen during takeoff or landing or during low-altitude flight.

Table 3.8 displays bird strike data for the UK and Canada (only) in different flight phases from 1990 to 2007 [2]. Table 3.8 presents the number of strikes during different flight phases and the corresponding percentage of strikes leading to damage at each flight phase.

Table 3.9 lists bird strike data from US civil aircraft (including US registered aircraft in foreign countries) at different flight phases over 26 years (1990–2015) [13]. Table 3.9 presents the number of strikes during different flight phases and the corresponding percentage of strikes leading to damage at each flight phase.

By comparing Tables 3.8 and 3.9, some agreements and differences will be noted. For example, the worst flight phase is landing for the UK and Canadian aircraft, while the approach phase is worst for US aircraft.

It is also important to clarify here that bird strike may differ by location. Two examples are described here. The first example is concerned with the Australian continent and the second is for Panama City.

Between 2004 and 2013, there were 14 571 birds strikes reported to the Australian Transport Safety Bureau (ATSB), most of which involved high capacity air transport aircraft [17].

Figure 3.22 shows the proportion of bird strikes in each phase of flight. Bird strikes reported during takeoff were most common (38%), followed by landing (36%), approach (18%), and initial climb (6%).

Figure 3.23 illustrates the percentages of bird strikes during different flight phases in Panama City's Tocumen International Airport from January 2013 to August 2015 [18].

Table 3.8 Bird strike data for the UK and Canada in different flight phases [2].

Flight phase	Number of strikes	Strike (%)	Damage (%)
Landing/Landing roll	1351	33	3
Approach	1130	28	7
Takeoff	996	24	5
Climb	433	11	10
Parked/Ground checks	53	1.3	13
En route	44	1	34
Taxi	30	0.7	3
Descent	30	0.7	10
Hover/Hover taxi/On deck	11	0.3	0
Total	4078	100	

Table 3.9 Reported bird strike with US civil aircraft (including US-registered aircraft in foreign countries) at different flight phases [13].

Phase of flight	Number struck	Total (%)
Approach	45 461	41
Takeoff run	19 800	18
Landing roll	18 862	17
Climb	18 828	17
Descent	3192	3
En route	3052	3
Departure	520	<1
Taxi	344	<1
Local[a]	298	<1
Arrival	155	<1
Parked	77	<1
Total known	110 589	100
Unknown[b]	53 855	
Total	164 444	

a) Phase of the flight was determined to be Arrival, Departure, or Local (i.e. pilot conducting "touch-and-go" operations) but the exact phase of flight could not be determined [13].
b) Of the 53 855 strike reports with "Unknown" phase of flight (all species), 39 387 (70%) were "Carcass Found" reports.

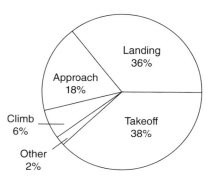

Figure 3.22 Statistics for bird strike in Australia for different phases of flight in the period 2004–2013.

The following conclusions may be made:

- The connection between the flight phase and altitude is evident.
- More than 80% of all bird strikes occur during the takeoff or landing phase of flight, which includes the takeoff roll, initial climb, final approach, and landing roll.
- A great number of strikes occur during the approach and landing flight phases because the flight path is limited due to other planes waiting to land and takeoff. Pilots also have to land correctly on the runway, which prohibits any maneuvering around birds.

FLIGHT PHASES WHEN BIRD STRIKES OCCUR AT TOCUMEN

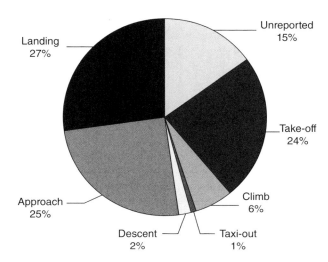

Figure 3.23 Statistics of bird strikes in Panama City's Tocumen International Airport from January 2013 to August 2015.

- Takeoff offers slightly more flexibility in flight path than landing but still has limited maneuverability.

3.7.5 Annual Bird Strike Statistics

Figure 3.24 illustrates the number of reported wildlife strikes with civil aircraft in the USA in the period 1990–2015. The total number of strikes is 166 276. Birds are

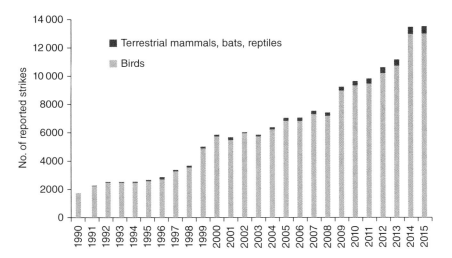

Figure 3.24 The number of reported wildlife strikes with civil aircraft, USA, 1990–2015. *Source:* courtesy USDA-APHIS [13].

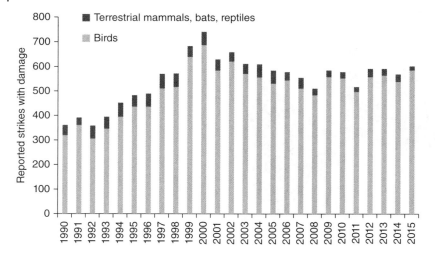

Figure 3.25 The number of reported wildlife strikes causing damage to civil aircraft, USA, 1990–2015. *Source:* courtesy USDA-APHIS [13].

responsible for 160 894 strikes while terrestrial mammals caused 3561 strikes. Bats are responsible for 1562 strikes. The smallest number of strikes (259) resulted from reptiles [13]. Annual numbers are continuously increasing from nearly 2000 strikes in 1990 to nearly 14 000 in 2015.

Figure 3.25 illustrates the number of reported wildlife strikes causing damage to civil aircraft, in the USA in the same period of 1990–2015. The total number of damaging strikes in these 26 years is 14 287, with birds being responsible for most of them (13 204) while terrestrial mammals caused 1073 accidents, bats resulted in only 8 accidents, and reptiles were responsible for 2 accidents [13]. The highest number of wildlife strikes was reported in the year 2000.

Figures 3.26 and 3.27 illustrate the strike rate and damage strike rate (per 100 000 aircraft movements) for two types of aircraft – namely commercial (air carrier, commuter, and air taxi service) and general aviation aircraft in the USA for 2000–2015 [13]. Bird strikes with US aircraft registered in foreign countries are excluded from this study.

Figure 3.28 displays the number of bird strikes in the USA, Canada, and the UK for 1990–2007 [2]. It shows an annual increase in the number of bird strikes of 3%.

Continuing with the annual bird strikes in the USA, Canada, and the UK for the period 1990–2007, analyses for the effect of bird mass (particularly those above 1.81 kg) were performed [2]. All strike reports for birds above 1.81 kg have been assembled into three groups:

- from 1.81 kg to below 3.6 kg;
- 3.6 kg (mainly the Canada goose);
- above 3.6 kg.

The 3.6 kg (Canada goose category) contributes 61% of all the strikes over 1.81 kg, while the other two categories show gentle growth. The number of strikes over 3.6 kg is small but their growth is of concern.

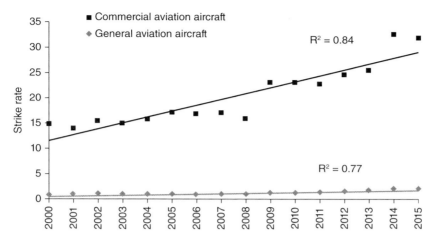

Figure 3.26 Strike rate (number of reported bird strikes per 100 000 aircraft movements) for commercial aviation aircraft and general aviation aircraft, USA, 2000–2015. *Source:* courtesy USDA-APHIS [13].

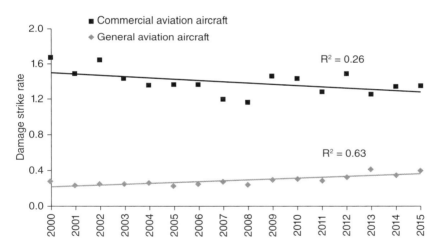

Figure 3.27 The damaging strike rate (number of reported damaging strikes per 100 000 aircraft movements) for commercial aviation aircraft and general aviation aircraft, USA, 2000–2015. *Source:* courtesy USDA-APHIS [13].

3.7.6 Monthly Bird Strike Statistics

Figure 3.29 demonstrates the percentage reported the monthly strikes of both birds and bats with civil aircraft in the USA, in the interval 1990–2015. Some 52% of bird strikes occurred between July and October. A maximum number of strikes occurred in August, while the least strikes occurred in January.

Concerning bats, 68% of strikes occurred between July and October. Maximum strikes occurred in August, while both January and December registered the minimum number of strikes.

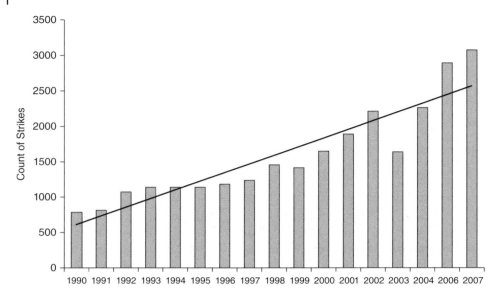

Figure 3.28 The number of bird strike in the USA, Canada, and the UK where the species is identified, 1990–2007.

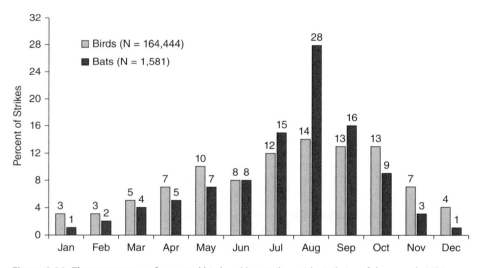

Figure 3.29 The percentage of reported bird and bat strikes with civil aircraft by month, USA, 1990–2015. *Source:* courtesy USDA-APHIS [14].

Figure 3.30 outlines that the monthly bird strike accidents in the period from 2002 to 2006 in the Australian continent [19]. The distribution of reported bird strikes by month and year is roughly bimodal with a first peak between March and May while the second peak is between October and January. For 2002–2004, this basic bimodal pattern applies. The second peak for the years 2002, 2005, and 2006 appear slightly earlier in October then falls away in November and rises again in December.

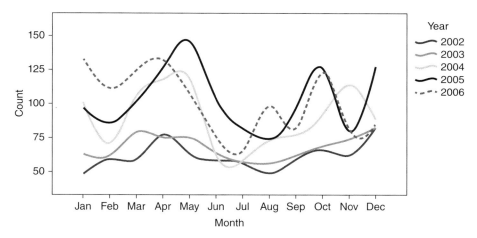

Figure 3.30 The number of bird and bat strikes with civil aircraft by month, Australia, 2002–2006.

3.7.7 Bird Strike by the Time of Day

Figure 3.31 shows that some 78% of bird strikes worldwide occur during the day [20]. Table 3.10 summarizes the reported time of occurrence of birds and bats strikes with civil aircraft in the USA over 26 years (1990–2015) [13].

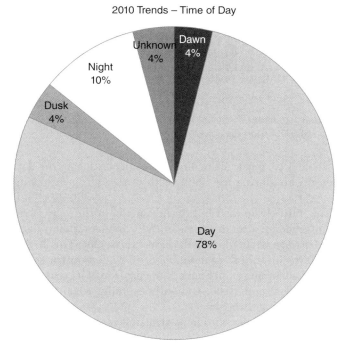

Figure 3.31 The number of bird strikes during different parts of the day and night. *Source:* courtesy Transport Canada [20].

Table 3.10 Reported time of occurrence of wildlife strikes with civil aircraft, USA, 1990–2015 [13].

Time of day	Birds		Bats	
	26-year total	Total known (%)	26-year total	Total known (%)
Dawn	3414	3	4	1
Day	64182	63	56	13
Dusk	4365	4	21	5
Night	30654	30	353	81
Total known	102615	100	434	100
Unknown	61829		1147	
Total	164444		1581	

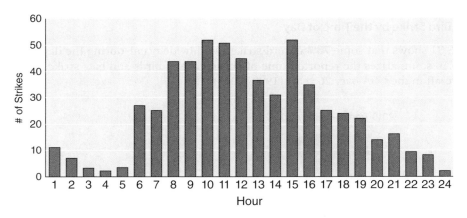

Figure 3.32 Daily bird strike distribution in Canada in 1999 (includes Canadian aircraft overseas and Canadian military aircraft). *Source:* courtesy Transport Canada [20].

From both Figure 3.31 and Table 3.10, most of the bird strikes occur during the day. The second largest percentage occurs during the night, with dusk and dawn having lower numbers.

Figure 3.32 illustrates the daily bird strike distribution in Canada in the year 1999. This distribution includes both Canadian aircraft overseas and Canadian military aircraft. The vast majority of bird strikes occur during daylight hours [20]. This is not surprising since fewer birds and fewer aircraft fly at night. Figure 3.32 demonstrates the substantial numbers of bird strikes occurring at all hours of the day. Small increases are evident in the morning, between 08:00 and 10:00, and early evening, 15:00 through 17:00, when the numbers of scheduled flights peaks.

Figure 3.33 illustrates the hourly bird strike record in Australia. It represents a different pattern from that of Canada. Bird strikes occurred most frequently during the busiest times of aircraft operation [19]. The busiest times are between 06:00 and 10:00 in the morning and 15:00 and 21:00 in the afternoon and evening.

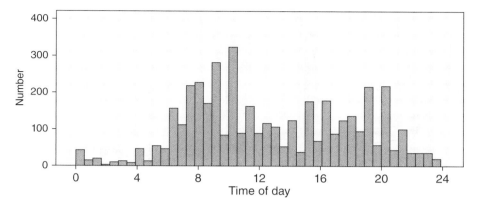

Figure 3.33 Number of bird strikes by the time of day in Australia (*N* = 4261).

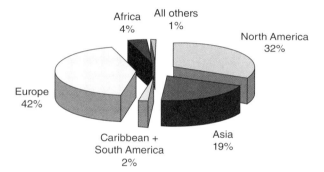

Figure 3.34 The percentage of bird strikes for different continents. *Source:* courtesy Airbus [9].

3.7.8 Bird Strike by Continent

Over 33 000 bird strike accidents to civil aircraft were recorded between 1990 and 2000 [9]. The percentage of these bird strikes for different continents is displayed in Figure 3.34. The highest percentage is recorded in Europe (42%), followed by North America (32%), next Asia (19%), then Africa (4%). The lowest figure is the Caribbean and South America (2%) [9].

Figure 3.35 illustrates nearly the same trends for bird strikes during the period 2008–2015 [8]. The total bird strike reports were 97 751 wildlife strikes. Europe's share was 35 775 strikes (36.6%), North America followed with 33 221 strikes (34%), Africa was in the third rank (instead of Asia in the 1990–2000 records) with a total number of 772 (0.8%).

3.7.9 Bird Strike by Weight of Birds

The distribution of the bird masses (in kg) involved in 33 accidents is presented in Figure 3.36 [2]. From the 33 accidents, 31 accidents were reported for birds heavier than 1 kg while the lightest bird mass recorded was 0.78 kg. The average mass for the birds over all the accidents was 3.8 kg.

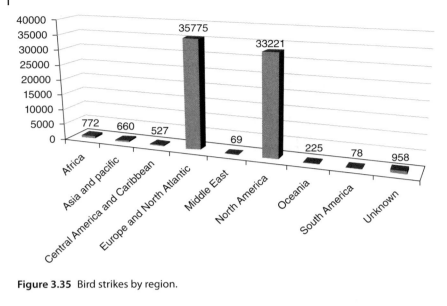

Figure 3.35 Bird strikes by region.

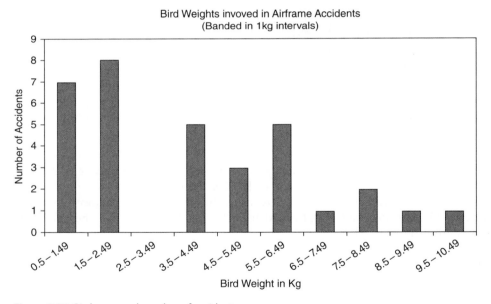

Figure 3.36 Bird mass and number of accidents.

3.7.10 Bird Strike by Aircraft Category

The total number of reports of bird strike in the period from 1990 to 2007 obtained from the UK (Civil Aviation Authority), Canada (Transport Canada), and the USA (US Department of Agriculture) is 94 743 [2]. There were only 10 919 strike reports involving airframes with complete data regarding both aircraft type and bird species

Table 3.11 Aircraft classification [2].

Aircraft category	CS category	Aircraft classification	No. of strikes	Damage (%)
1	CS-23	Normal/Utility/Aerobatic (Propeller driven)	1369	34.6
2	CS-23	Normal/Utility/Aerobatic (Jet driven)	72	29.2
3	CS-23	Commuter (both propeller and jet driven)	418	27.5
4	CS-23	Business Jets (both propeller and jet driven)	226	26.6
5	CS-25	Large Aircraft (Propeller driven)	1375	8.7
6	CS-25	Large Aircraft (Jet driven)	7266	9.3
7	CS-27	Small helicopters	65	49.2
8	CS-29	Large helicopters	128	14.1

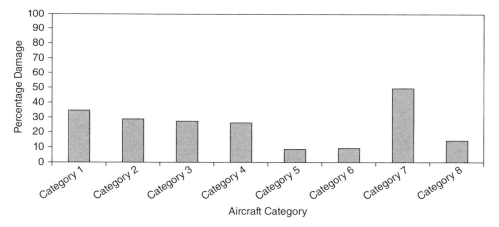

Figure 3.37 Aircraft category plotted against the percentage of strikes causing damage.

hit. Moreover, of these 10 919 strikes, only 1517 (13.9%) resulted in some level of damage to the aircraft.

Table 3.11 and Figure 3.37 shows that there is a direct relationship between the type of aircraft and the number of reported strikes resulting in damage. Those aircraft which are the subject of comprehensive bird strike certification requirements (large aircraft driven either by turboprop or turbofan engines, Categories 5 and 6, and large helicopters, Category 8) are much less likely to sustain damage. These results may be because transport aircraft pilots may be more likely to report bird strikes that do not result in damage.

The worst case is that of small helicopters (Category 7). This is based on the smallest number of reports (65) for all categories but may indicate a particular risk for this category of aircraft.

Statistics for multiple bird strikes are given in Table 3.12 and Figure 3.38. They are arranged in the three groups of birds (single, small number ranging from 2 to 10, and

Table 3.12 Percentage of multiple strikes causing damage to each aircraft category [2].

Aircraft category	Number of birds	No. of Strikes	Damage (%)
1	1	1011	32.9
	2–10	320	38.8
	>10	22	45.5
2	1	44	22.7
	2–10	25	44
	>10	2	—
3	1	278	26.3
	2–10	126	28.6
	>10	9	55.6
4	1	139	23
	2–10	75	32
	>10	12	33.3
5	1	973	6.2
	2–10	327	14.4
	>10	42	29.3
6	1	5083	7.1
	2–10	1842	15.5
	>10	234	19.2
7	1	58	48.3
	2–10	7	57.1
	>10	0	—
8	1	118	13.6
	2–10	4	25
	>10	0	—

flock of birds greater than 10) striking different categories of aircraft and the percentage of damage in each case.

Category 1, representing the small propeller aircraft, and Category 6, representing the large jet airplanes, are subject to the highest number of bird strikes.

3.7.11 Bird Strike by Bird Species

In order to estimate the frequency of strikes with a particular species in a given year, a mean value for the total number of strikes in the last five years with each bird species is calculated. Such a five-year mean ensures that the probability estimate is based on relatively recent data.

A risk assessment matrix, as developed by Allan [21], involves the generation of a simple probability/severity matrix. Both the probability and the severity measures may be split into five categories, as described in Table 3.13.

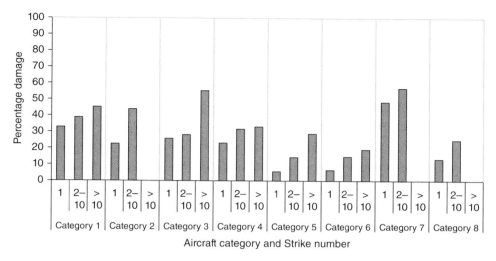

Figure 3.38 The proportion of multiple strikes resulting in damage by aircraft category.

Table 3.13 Severity and probability levels for a strike with a particular bird species.

Number of strikes per year	>10	3–10	1–2.9	0.3–0.9	0–0.2
Probability category	Very high	High	Moderate	Low	Very low
Percentage of strikes causing damage for a bird species in a certain period, national data 1975–1995 [21]	>20	10–20	6–9.9	2–5.9	0–1.9
Severity category	Very high	High	Moderate	Low	Very low

"Probability" is defined here as the number of strikes per year (5-year rolling mean from airport data). The term "severity" is defined as the percentage of strikes causing damage for a bird species based on the national database (1975–1995).

The categories defined in Table 3.13 are used in the UK based on the field experience of airport bird control staff and managers, as well as 30 years of research experience with the staff of Central Science. Individual airports could set their own category boundaries.

Figure 3.39 illustrates the severity versus the probability of bird strike for different species in the UK [21]. Species having a very high severity are the great cormorant (mass 2.6–3.7 kg), though its strike probability is very low. On the contrary, swallow, martin, swift larks, and pipets species have high probabilities of bird strike but very low severity. Strikes with wood pigeon species have moderate probability and moderate severity.

A list for all the critical bird species which frequently collide with aircraft causing damage or having a negative effect-on-flight (EOF) in the USA in the period 1990–2015 is listed in very lengthy tables [13]. Only a sample of these reports is given here in Table 3.14.

In another study, the top five species causing moderate and substantial damage in the US (1990–2007) have been identified [2] and are listed in Table 3.15. As stated

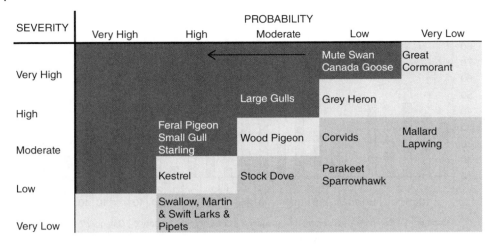

Figure 3.39 Severity versus the probability of bird strike for different species.

Table 3.14 Samples for critical bird species that frequently collide with aircraft causing damage or having a negative effect-on-flight (EOF) in USA (1990–2015) [13].

Bird species	Total number of strikes	Strikes with damage	Strikes with negative EOF	Strikes with multiple birds	Aircraft downtime (hours)	Costs (US$)
Waterfowl (Canada goose, snow goose, Mallard)	4951	2011	1043	1710	162 974	243 843 168
Hawks, eagles, vultures	5938	1428	936	190	134 967	111 946 408
Gulls	10 586	1446	1204	2188	60 005	58 887 723
Pigeons, doves	12 196	497	643	2373	28 479	22 354 509
Starlings, mynas	4029	131	190	1373	3113	7 127 433
Total known birds	86 543	7012	5597	14 489	455 640	533 963 041
Total unknown birds	77 901	6546	4288	7747	176 721	132 130 177

previously, 94 743 strikes were reported during this period. However, complete data were available for only 10 919 strikes (recording both aircraft type and bird species). Moreover, of these 10 919 strikes, only 1517 (13.9%) resulted in some level of damage to the aircraft. Most of the reports (77%) record no damage, 16% are recorded as moderate, and only 6% are recorded as substantial.

3.7.12 Populations of Some Dangerous Bird Species in North America

The increase in populations of hazardous wildlife species continue to challenge airports' ability to provide a safe operating environment. Numbers of American white

Table 3.15 Top five species causing moderate/substantial damage from 1990 to 2007 (the USA only) [2].

Damage type	Species	Damage (%)	Number of strikes
Moderate	Canada goose	29.1	313
	Rock pigeon	7.5	81
	Turkey vulture	7.3	78
	Red-tailed hawk	6.3	68
	Mallard	4.7	51
Substantial	Canada goose	31.8	135
	Turkey vulture	9.2	39
	Rock pigeon	7.1	30
	Mallard	5.2	22
	Snow goose	4.2	18

Figure 3.40 Rapid increase in the population of American white pelican. *Source:* courtesy USDA-APHIS [22].

pelicans increased six-fold [22] between around 1990 and 2009, as illustrated in Figure 3.40.

Growth in the geese population, and especially the increase in non-migratory geese near urban centers, is causing considerable air safety concern.

Figure 3.41 illustrates the increase in the number of snow geese in North America from 2.0 million in 1980 to 6.8 million in 2013 [13].

Numbers of Canada geese (Figure 3.41) have ballooned to more than 3.5 million. The North American population of Canada geese has stabilized in recent years after a period

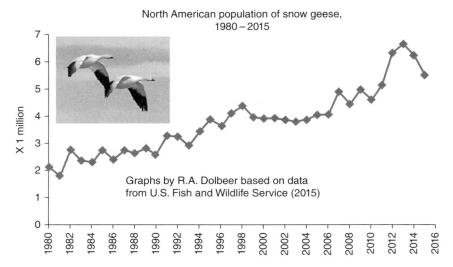

North American population of snow geese, 1980 – 2015

Graphs by R.A. Dolbeer based on data from U.S. Fish and Wildlife Service (2015)

Figure 3.41 Population of North American snow geese. *Source:* courtesy USDA-APHIS [13].

Figure 3.42 A Canada goose. *Source:* courtesy USDA-APHIS [14].

of rapid expansion of the resident (non-migratory) population from 1984 to 2000, as illustrated in Figure 3.43.

3.7.13 Dangerous Bird Species in Europe

Gulls and diurnal raptors represent the greatest danger for both military and civil aircraft in Europe [23]. Swallows, swifts, pigeons, European starlings, Eurasian kestrels, and northern lapwings are less dangerous.

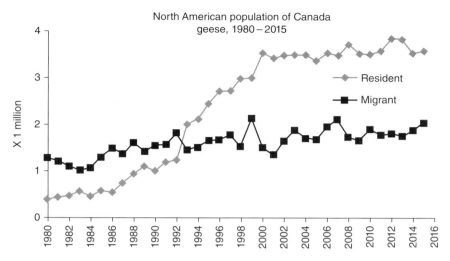

Figure 3.43 Population of North American Canada geese. *Source:* courtesy USDA-APHIS [13].

The following species are the most dangerous in each country or part of Europe:

- gulls in the British Isles (UK and Ireland) and in the coastal parts of the continent;
- swans, common gulls, and herring gulls in northern parts of Europe (Norway, Sweden, and Finland);
- yellow-legged gulls and vultures in southern Europe (Bulgaria, Greece, Croatia, Italy, and Spain);
- diurnal raptors (kestrels, buzzards), corvids, storks, and gulls in western Europe (Belgium, Netherlands, and Germany) and central Europe (former Czechoslovakia, Poland, and Hungary as well as Lithuania and Estonia).

3.8 Military Aviation

3.8.1 Introduction

Databases for bird strikes with military aircraft are not as complete as for civilian aircraft. Generally, reports that describe and analyze bird strikes in military aviation are much less detailed than those describing civil aviation [24].

In a similar way to the ICAO in its annex 13 for civilian aircraft, the United States Air Force (USAF) adopts the following definitions for accident and incident [25] due to bird strike.

- Accident: if the aircraft damage requires significant man-hours to repair or results in injury or death to the crew, it is denoted an accident.
- Incident: if the aircraft damage requires some repair before the next flight, or if it constitutes a "significant hazard to the crew or aircraft," it is identified as an incident.

It is also important to clarify that military operations differ from civilian ones mainly in their requirement for high-speed, low-level flying. This type of flight activity has accounted for the greatest losses in the US for both aircraft and aircrew lives [25].

The following statistics for accidents caused by bird strikes with military aircraft cover 1950 to 1999 [26]:

- a total of 286 serious accidents to military aircraft from 32 countries;
- most of these countries are in Europe (east to Russia), plus Canada, USA, Australia, and New Zealand;
- of these 286 accidents, at least 63 were fatal, with at least 141 deaths (137 aircrew and 4 on the ground);
- the costliest decade was the1990s, with at least 68 bird-related fatalities;
- countries with the highest known numbers of bird-related accidents are Germany (60 aircraft from at least eight countries), the UK (47), and the USA (46+);
- the most serious accidents involve jet fighter or attack aircraft with one engine (at least 179 accidents), with two engines (40+), and jet trainers (34+);
- only seven four-engined heavy transport military aircraft have been lost since 1950 (three in the 1990s);
- some accidents to military aircraft were reported in Asia (especially India);
- few accidents were reported in Africa and South America;
- a detailed description of these 286 accidents is given by McCracken [25].

Lists and analysis of bird strike accident encountered by the USAF during the years 1966 to 1975 have been arranged and stored by McCracken [25].

Most bird strikes caused no damage to military aircraft. For example, in 32 countries during the period January 1985 to February 1998, more than 95% of strikes (33 262 of 34 856) were reported to be non-damaging, incurring less than US$ 10 000 in damage per occurrence [26]. Damaging strikes are categorized by cost into three classes, as follows:

- Class C – repair cost ranges from US $10 000 to US$ 200 000;
- Class B – repair cost ranges from US$ 200 000 to US$ 1 million;
- Class A– costs are either equal to or more than US$ 1 million in damage or involve the loss of an aircraft and a fatality.

It is noticeable that statistics for bird strikes with military aircraft in the last 10 years are regional and not worldwide. Statistics cover a country or a region of the world rather than the whole continents.

For example, a statistic from the USAF for 2010 is reported by Nicholson and Reed [11] and states that 5000 bird strikes were reported.

Another study dealt with bird strikes with military aircraft in Poland in the period November 2011 to the end of September 2016 [27]. In total, 141 bird strikes were recorded, including 26 collisions which resulted in damage to aircraft assemblies. During this period, bird strikes did not cause any aircraft accidents. Most of the bird strikes were with F-16 (27), Hercules C-130 (19), and MiG-29 (17) aircraft.

3.8.2   Annual Bird Strike with Military Aircraft

Figure 3.44 illustrates the annual number of incidents (dashed line) and accidents (solid line) caused by bird strikes in the 10 years 1966 to 1975. The annual number of incidents remained between 300 and 400. Bird strikes in 1974 were high, with 467, but 1975 was close to normal with 402 strikes [25].

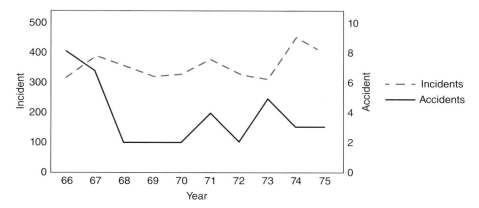

Figure 3.44 Military accidents and incidents in the period 1966–1975. *Source:* courtesy IBSC [24].

Accidents in all categories are represented by the solid line in Figure 3.44. Since 1965 there have never been fewer than two bird strike accidents per year, and since 1969 the average has been three per year. Over these 10 years, these accidents have resulted in the loss of 14 aircraft and seven pilots [25]. Bird strikes were also strongly suspected in several other accidents involving aircrew fatalities and destroyed aircraft.

Another study, covering the period 1990–2013, concluded that the number of strikes with military aircraft is only 10% or even less than those associated with civilian aircraft [28]. Figure 3.45 illustrates such a comparison.

The most recent study was carried out by *Military Times*, which published a searchable database for military aviation mishap reports for the fiscal years 2011–2017 [29]. The database includes more than 7500 individual records for all Class A,

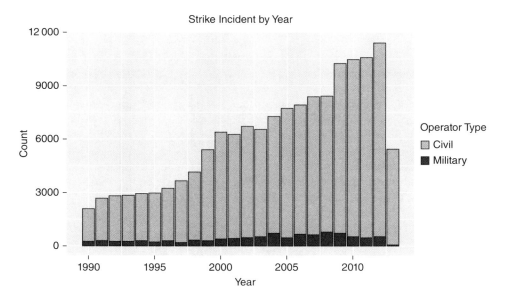

Figure 3.45 Bird strike counts for civilian and military aircraft in the period 1990–2013.

Table 3.16 Class A, B, and C aviation mishaps due to bird strike.

	Year						
	2011	2012	2013	2104	2015	2016	2017
Air force	46–50 per year			64	74	67	71
Navy	10	9	12	15	9	17	15

B, and C aviation mishaps. This database can be searched by aircraft type, base, fiscal year, and location [29].

About 8% of all Class A, B, and C aviation mishaps in the period 2011–2017 resulted from wildlife strikes, according to data obtained by *Military Times* [30]. Two valuable findings from this study for the period 2011–2017 are as follows.

- Looking at air force aircraft, some 418 wildlife strikes were reported out of the 5109 Class A, B, and C mishaps. The corresponding costs were nearly US$ 182 million, based on Bird/Wildlife Aircraft Strike Hazard team.
- Naval aircraft encountered 87 Class A, B, and C wildlife strikes out of a total of 1076 mishaps during that time frame. Based on the average cost per event, *Military Times* estimates the damage at roughly US$ 64.8 million.

The detailed annual strike data are shown in Table 3.16 for both air force and naval aircraft.

3.8.3 Annual Costs of Bird Strike with Military Aircraft

Figure 3.46 displays the annual cost in US$ due to bird strikes with military aircraft in the US in the period from 1965 to 1975. Over US$ 81 million was lost to damaged or

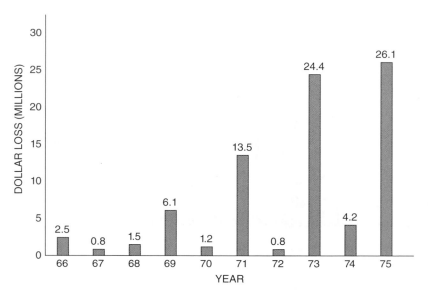

Figure 3.46 Annual cost of bird strike from 1966 to 1975. *Source:* courtesy IBSC [24].

destroyed aircraft. From 1970 to 1975, approximately US$ 70 million has been lost, of which over US$ 61 million involved the cost of destroyed aircraft. However, these figures do not include any cost for the time required to repair the damaged aircraft. Consequently, actual costs are necessarily much higher.

The costs for bird strikes for the period 2011–2017, as reported by *Military Times* [30] are:

- air Force aircraft cost are nearly US$ 182 million, based on Bird/Wildlife Aircraft Strike Hazard team;
- naval aircraft cost are estimated at nearly US$ 64.8 million.

3.8.4 Statistics of Bird Strike by Altitude

From 1970 to 1980, 80% of bird strikes occurred at altitudes from 0 to 3000 ft above ground level, with 37% occurring below 300 ft [31]. This is due to two factors:

- birds routinely fly at these altitudes (except when migrating);
- aircraft must pass through these altitudes when they take off and land.

Figure 3.47 shows bird strikes by impact altitude up to 3000 ft. It shows that over 50% of bird strikes occurred below 3000 ft (data for 1975). Moreover, some 32% of strikes occur at altitudes less than 500 ft. The high percentage of the unknown altitudes (41%) results from the number of bird strikes which were not noticed until after the flight or for which the crew failed to report the altitude.

Table 3.17 categorizes bird strikes with military aircraft against the impact altitude during the period from 1970 to 1980 [31]. Table 3.17 shows high numbers of very low altitude strikes. For altitudes 0–500 ft, the 1970–1980 bird strikes represent 61%, which is nearly double that for 1975 (32%).

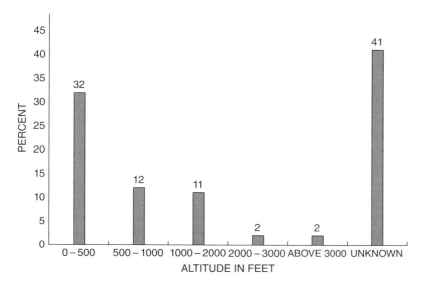

Figure 3.47 Bird strikes with military aircraft in 1975 versus flying altitude. *Source:* courtesy IBSC [24].

Table 3.17 Bird strikes by impact altitude [31].

Altitude (ft)	0–500	500–1000	1000–1500	1500–2000	2000–2500	2500–3000	>3000
Bird strike (%)	61.2	16.1	8.2	8.0	1.6	1.8	3.1

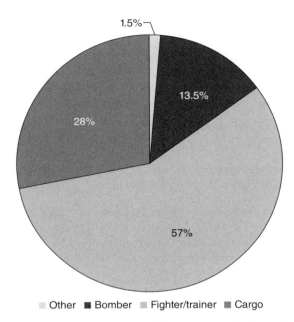

Other ■ Bomber ▧ Fighter/trainer ▨ Cargo

Figure 3.48 Bird strikes with different types of military aircraft. *Source:* courtesy IBSC [24].

3.8.5 Bird Strike by Aircraft Type

Figure 3.48 displays the percentages of bird strike with different types of aircraft in the period 1966–1975. Fighters and trainers were subjected to the highest percentage (57%) since they spend a great deal of time in the airdrome environment practicing takeoffs and landings. Next comes cargo aircraft (28%) with bombers being subject to the smallest number of strikes (13.5%).

Table 3.18 categorizes bird strikes by military aircraft type during the period from 1970 to 1980 [31]. The statistics are very similar. Fighters and trainers are combined in reference [25] and separated in reference [31]. Fighters and trainers represent the mostly collided types (57% or 61.3%). Cargo aircraft are in second place, with 28% in both studies. Bombers follow with 13.5% or 7.9%.

Table 3.18 Bird strikes by different military aircraft groups [31].

Aircraft group	Fighter	Cargo	Trainer	Bomber	Helicopter
Bird strike (%)	42.2	28.4	19.1	7.9	2.2

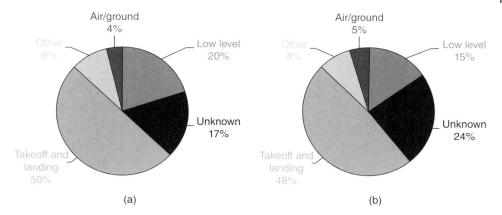

Figure 3.49 Bird strike at different flight phases for military aircraft. (a) In 1974, (b) in 1975. *Source: courtesy IBSC [24].*

Table 3.19 Bird strikes for military aircraft by flight phase [31].

Flight phase	Takeoff	Climb	Cruise	Descent	Range (munitions firing)	Low level	Unknown	Final approach	Landing
Bird strike (%)	17.9	1.9	4.7	0.98	7.1	14.6	28.79	9.42	13.0

3.8.6 Bird Strike by Flight Phase

Figure 3.49 displays the percentages of bird strike with military aircraft at different flight phases in 1974 and 1975.

In 1974, half of the bird strikes occurred during the takeoff or landing phase [25] and 20% in the low-level phase (Figure 3.49a). These flight phases represent a significant hazard to flight safety because of the high speeds and low altitudes involved.

In 1975, 48% of bird strikes occurred during the takeoff or landing phase [25] and 15% in the low-level phase (Figure 3.49b).

Other statistics for bird strikes with military aircraft are listed in Table 3.19 [31].

3.8.7 Bird Strike by the Distance from the Base

Most of the bird strikes for military aircraft occur over or close to the air base. Figure 3.50a shows that the percentages of bird strikes for military aircraft in 1974 were categorized as follows:

- 39% over the base;
- 8% within 10 miles from the base;
- 9% for distances of 10 miles or more.

The large unknown category (44%) shown in Figure 3.50a include the low-level and initial climb out bird strikes in which the nearest base is unknown.

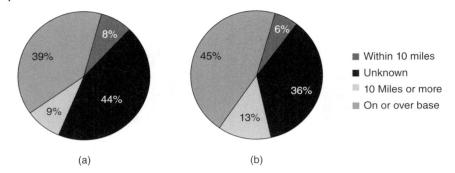

Figure 3.50 Bird strikes by the distance from the base. (a) In 1974, (b) in 1975. *Source:* courtesy IBSC [24].

Figure 3.50b shows that the percentages of bird strikes for military aircraft in 1975 were categorized as follows:

- 45% on or over the base;
- 6% within 10 miles from the base
- 13% for distances of 10 miles or more.

Again, the large unknown category (36%) shown in Figure 3.50b inclues the low-level and initial climb out bird strikes in which the nearest base is unknown.

Statistics for the period 1970–1980 outlined that 63% of all bird strikes occur around airfields, with 37% within 10 miles of an airfield [31].

3.8.8   Bird Strike by Month

During spring and fall (autumn), bird migration leads to a significant increase in bird strikes for military aircraft. Figure 3.51 shows an increase of bird strikes with military aircraft in April and May due to the spring bird migration. Moreover, since bird's fall migration starts in August, heavy strikes are taken again in September, October, and November. Waterfowl, passerines, and shorebirds are responsible for the major damaging strikes during migratory periods. These strikes usually result in damage to the airframe from dents on leading edges of wings, tail assemblies, and the nose of the aircraft.

3.8.9   Bird Strike by the Time of Day

Table 3.20 summarizes the time of occurrence of bird strikes encountered in 1983 for US military aircraft. 67% of strikes occurred during daylight versus 18% during the night [31]. Dawn and dusk strikes make up only 5%. Unknown strike times were only 10% of the total.

3.8.10   Bird Strike by Part

Table 3.21 lists the percentages of the parts of military aircraft struck by birds [31].

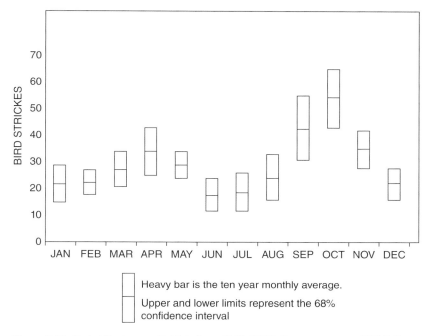

Figure 3.51 Bird strike by month (data from 1965–1975). *Source:* courtesy IBSC [24].

Table 3.20 Reported time of occurrence of bird strikes with military aircraft, USA, 1983 [31].

Time of day	Percentage
Dawn/Dusk	5
Day	67
Night	18
Total known	90
Unknown	10
Total	100

Table 3.21 Percentage of bird strike of military aircraft by part [31].

Aircraft part	Engine/ engine cowling	Windshield	Wings	Radome/ nose	Fuselage	External tanks/Pods/ Landing gear	Multiple hits	Other
Bird strike (%)	22.3	20.6	19.3	15.1	8.9	6.7	5.2	1.9

Table 3.22 Costs of bird strikes with US Air Force and Navy aircraft (2011–2017) [31].

Bird species	Costs (US$)
Snow goose	9 400 918
Mallard	10 304 474
Red-tailed hawk	17 039 479
Spot-billed duck	24 955 020
Mourning dove	29 290 414
Turkey vulture	37 767 636
American white pelican	41 760 459
Pink-footed goose	43 262 092
Black vulture	75 686 764
Canada goose	93 812 397

3.8.11 Critical Bird Species

The cost of bird strikes with the US Air Force and Navy aircraft depends on bird species. Table 3.22 lists the costs associated with different species [31].

3.9 Bird Strikes on Helicopters (Rotating Wing Aircraft)

All rotating wing aircraft (helicopters) are powered by only shaft-engines including both piston engines and turboshaft engines [32].

3.9.1 Bird Strike with Civilian Helicopters

Helicopters normally fly at low altitudes and thus could be surrounded by lots of birds. Consequently, they are subjected to numerous bird strikes. In many cases, birds penetrate the windshield, potentially leading to pilot incapacitation that could cause fatalities for everybody aboard.

Bird strikes with helicopters have increased dramatically in recent years. Following the FAA, the collision of birds with a helicopter can be categorized as:

- accidents associated with fatalities;
- incidents that damage the aircraft and create the potential for crashes.

Figure 3.52 illustrates the number of bird strikes with helicopters in the period 1990–2013. A rapid increase is shown:

- only 14 cases were reported in 1990;
- 34 cases were reported in 1999;
- 136 cases were reported in 2009
- the number had increased to 213 cases in 2013 [33].

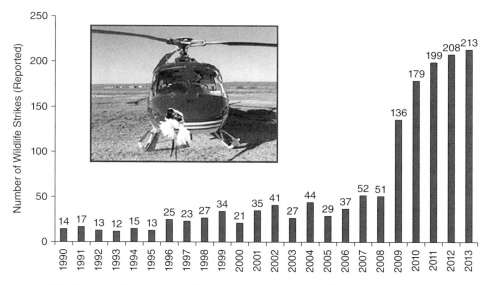

Figure 3.52 Number of birds strikes with helicopters in the period 1990–2013. *Source:* courtesy USDA-APHIS [33].

This increase is due to the following:

- greater awareness among pilots about the importance of reporting bird strikes since the famous accident involving the ditching of US Airways Flight 1549 in New York's Hudson River in 2009;
- a rise in populations of large bird species in North America, including the American white pelican, the American snow goose, and the Canadian goose – other large-bird species with rising populations in the world include bald eagles, wild turkeys, turkey vultures, double-crested cormorants, sandhill cranes, great blue herons, and ospreys.

Current FAA regulations for helicopters weighing more than 7000 lb state that windshields must withstand an impact with a 2.2 lb bird. However, no bird strike safety standards exist for helicopters weighing less than 7000 lb, which represents about 90% of the US fleet, including all tour and medical helicopters. For this reason, the staff of FAA's helicopter directorate is urging that a special industry committee is established with government backing to examine the following:

- should there be changes in the standards for helicopter construction and operation for better protection against bird strikes;
- the necessity for new technology to quickly disperse birds in the way of helicopters, possibly using strobe lights.

The critical parts of the helicopter which are subjected to most bird strikes are the windscreen, main rotor, radome/nose, fuselage, tail, landing gear, and lights. Figure 3.53 illustrates the data about helicopter parts struck by birds. The windscreen is the most critical part (33.8%), followed by the main rotor (17.9%), and then the radome/nose (13.6%).

Main Rotor System
17.9%

Multiple locations
23.5%

Tail Rotor
1.5%

Windscreen
33.8%

Fuselage
9.7%

Radome Nose
13.6%

Figure 3.53 Percentages of bird strike with different parts of a helicopter [38].

An example of a helicopter bird strike accident may be described as follows. A pilot in the Gulf Coast region was flying at about 1000 ft and 115 mph when two ducks slammed through the windshield and hit him in the face. The pilot had so much bird gore on his face that he could not immediately breathe or see. However, he managed to land the helicopter safely without injuring any of the other five people on board.

3.9.2 Bird Strike with Military Helicopters

A detailed study looking at bird strikes with military helicopters in the USA ended with a conclusion that helicopters operated by the US Army, Navy, and Air Force are subjected to the biggest threat [22].

The number of wildlife strikes on helicopters across all branches of the US military (Army, Navy, Air Force, and Coast Guard) from 1979 to 2011 reached 2511. Birds are particularly problematic to both the Apache attack helicopters and huge Chinook helicopters that transport troops, supplies, and artillery to and from the battlefield. Though wildlife strikes with helicopters occurred in almost every state, Florida had the highest number of incidents (617 strikes), followed by New Mexico (204 strikes), and Georgia (192 strikes).

Based on the military's records, wildlife strikes caused eight injuries – mostly cuts, lacerations, or bruising when birds crashed through the windscreen of the aircraft (data from 1993 to 2008) – and two deaths in the United States so far [34].

Flying creatures are responsible for 812 out of the 2511 incidents recorded for military helicopter incidents. Birds were the culprits in 91% of these cases while bats were responsible for the remaining 9%.

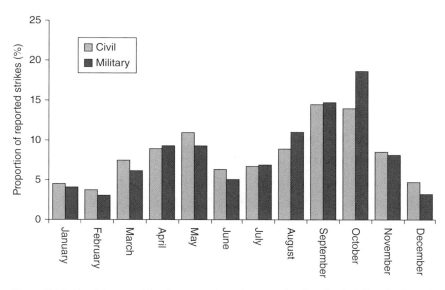

Figure 3.54 Monthly record for the proportion of reported strikes, both civil and military helicopters. *Source:* courtesy IBSC [24].

Concerning bird species striking helicopters, it was found that:

- Air Force helicopters were commonly struck by warblers (16.8%) and perching birds (12%);
- Naval vehicles tended to be hit by gulls (18.2%), seabirds (14.9%), shorebirds (13.4%), and raptors and vultures (12.6%).

Concerning monthly bird strikes, it was found that almost 42% of the recorded wild-life strikes occurred between September and November, making that period the most prevalent for the accidental collisions (Figure 3.54). The months of December and February were less hazardous, with 10.4% of wildlife strikes occurring in those months.

The cost of these accidents ranges from US$ 12 000 to US$ 337 000.

In some cases, wildlife strikes can be fatal for those onboard the helicopter. An example of a fatal accident is that of an HH-60G helicopter which was struck by a pink-footed goose on 8 January 2014 (Figure 3.55). During a flying exercise the helicopter crashed in Cley Marshes, Norfolk, UK, resulting in four fatalities [23].

3.10   Birds Killed in Strikes with Aircraft

Thousands of accidents/incidents involving planes and birds happen every year. The FAA database counts such annual accidents. If at least one bird is killed in each strike, then the number of birds killed by strikes with aircraft is greater than or equal to the number of such incidents [35]. The FAA counted 13 159 incidents for civilian aircraft in 2014, so at least this number of birds was killed in that year. Since less than 10% of bird strikes are normally encountered by military aircraft, then some 10% may be added to the above figure for the fiscal year 2014, to reach nearly 15 000 killed birds.

Figure 3.55 Crash of HH-60G helicopter due to strike with a Pink-footed goose. *Source:* courtesy IBSC [24].

The assumption that only one bird is killed in each strike is a very modest number. For example, an F16 aircraft leaving its base in Mazer-e-Sherif, Afghanistan on 23 October 2012 resulted in 40 dead birds after encountering a flock of 99 small birds.

References

1 Greenspan, J. (2015). Everything You Need to Know About Birds and Planes. *Audubon Magazine.* https://www.audubon.org/news/everything-you-need-know-about-birds-and-planes (accessed 14 January 2019).

2 Dennis, N. and Lyle, D. (2009). Bird Strike Damage & Windshield Bird Strike, Final Report, ATKINS Final Report, 5078609-rep-03, Version 1.1 https://www.easa.europa.eu/sites/default/files/dfu/Final%20report%20Bird%20Strike%20Study.pdf (accessed 2 January 2019).

3 DeFusco, R.P. and Unangst, E.T. (2013). Airport Wildlife Population Management, A Synthesis of Airport Practice, ACRP Synthesis Report 39. https://doi.org/10.17226/22599.

4 Satheesan, S.M. (1998). Need For Imparting Training At National Level To Bird Controllers At Civil And Military Airports, IBSC 24/WP1, Slovakia, 14–18 September 1998.

5 Maragakis, I. (2009). European Aviation Safety Agency, Safety Analysis and Research Department Executive Directorate, Bird Population Trends and Their Impact on Aviation Safety 1999–2008.

6 Thorpe, J. (2012). 100 Years of Fatalities and Destroyed Civil Aircraft Due to Bird Strike. IBSC30/WP Stavanger, Norway 25th–29th June 2012. http://www.int-birdstrike.org/Warsaw_Papers/IBSC26%20WPSA1.pdf (accessed 2 January 2019).

7 Elsayed, A.F. (2017). *Aircraft Propulsion and Gas Turbine Engines.* 2nd edition, Boca Raton: Taylor & Francis, CRC Press.

8 International Civil Aviation Organization (2017). Summary of Wildlife Strikes Reported to the ICAO Bird Strike Information System (IBIS) for the Years 2008–2015. Electronic Bulletin, EB 2017/25, Attachment B. https://www.icao.int/safety/IBIS/2008%20-%202015%20Wildlife%20Strike%20Analyses%20(IBIS)%20-%20EN.pdf (accessed 2 January 2019).

9 AIRBUS (2004). Flight Operation Briefing Notes, Operating Environment: Birdstrike Threat Awareness, Section IV (Operational Effects of Birdstrikes), part 1 (General). http://www.skybrary.aero/bookshelf/books/181.pdf (accessed 2 January 2019).

10 Reddy, G. (2012). Bird Strike. https://www.slideshare.net/gyanireddy/bird-strike-11316202 (accessed 2 January 2019).

11 Nicholson, R. and Reed, W.S. (2011). Strategies for Prevention of Bird-Strike Events, *AERO* QTR_03.11, Boeing. http://www.boeing.com/commercial/aeromagazine/articles/2011_q3/4/ (accessed 2 January 2019).

12 Aircraft Damages Caused by Birds, Aviation Ornithology Group, Moscow. http://www.otpugivanie.narod.ru/damage/eng.html (accessed 2 January 2019).

13 Dolbeer, R.A., Weller, J.R., Anderson, A.L., and Begier, M.J. (2016). Wildlife Strikes to Civil Aircraft in the United States, 1990–2015, FAA report, November 2016. https://wildlife.faa.gov/downloads/Wildlife-Strike-Report-1990-2015.pdf (accessed 2 January 2019).

14 Dolbeer, R.A. and Barnes, W.J. (2017). Positive bias in bird strikes to engines on left side of aircraft. *Human–Wildlife Interactions* 11 (1): 33–40.

15 Clearly, E.C., Dolbeer, R.A., and Wright, S.E. (2003). Wildlife Strikes to Civil Aircraft in the United States, 1990–2002. FAA Report.

16 Clearly, E.C. and Dolbeer, R.A. (2005). *Wildlife Hazard Management at Airports: A Manual for Airport Personnel*, 2e. https://digitalcommons.unl.edu/cgi/viewcontent.cgi?article=1127&context=icwdm_usdanwrc (accessed 2 January 2019).

17 Australian Transport Safety Bureau (2014). ATSB Transport Safety Report 2014–15. https://www.atsb.gov.au/media/5366635/ATSB%20Annual%20Report%202014%C2%AD15.pdf (accessed 2 January 2019).

18 Cheng, E., Hlavenka, T., and Nichols, K. (2015). It's A Bird, It's A Plane, It's A Problem: An Analysis of Bird Strike Prevention Methods at Panama City's Tocumen International Airport. https://web.wpi.edu/Pubs/E-project/Available/E-project-102915-205833/unrestricted/CopaBirdStrikePreventionMethodsInPanama.pdf (accessed 2 January 2019).

19 Stanton, D.R. (2008). An analysis of Australian birdstrike occurrences 2002 to 2006. Australian Transport Safety Bureau, Report AR-2008-027. https://www.atsb.gov.au/media/27782/ar2008027.pdf (accessed 2 January 2019).

20 MacKinnon, B. (2004). Sharing the Skies, Chapter 7, Bird- and Mammal-strike Statistics. https://www.tc.gc.ca/eng/civilaviation/publications/tp13549-chapter7-2144.htm (accessed 2 January 2019).

21 Allan, J. (2006). A heuristic risk assessment technique for birdstrike management at airports. *Risk Analysis* 26 (3), Wiley online library, https://doi.org/10.1111/j.1539-6924.2006.00776.x.

22 Dolbeer, R.A., Wright, S.E., Weller, J. and Begier, M.J. (2012). Wildlife Strikes to Civil Aircraft in the United States, 1990–2010. Serial Report Number 17. FAA and USDA-APHIS.

23 Kitowski, I. (2011). Civil and military birdstrikes in Europe: an ornithological approach. *Journal of Applied Sciences* 11: 183–191.

24 Richardson, W.J. and West, T. (2000). Serious Birdstrike Accidents to Military Aircraft: Updated List and Summary, International Bird Strike Committee. IBSC 25/WP SA1. Published in IBSC Proceedings: Papers & Abstr. 25 (Amsterdam, vol. 1): 67-97(WP SA1). http://worldbirdstrike.com/IBSC/Amsterdam/IBSC25%20WPSA1.pdf (accessed 2 January 2019).

25 McCracken, P.R. (1976). Bird Strikes And The Air Force. Bird Control Seminars Proceedings, Paper 53. http://digitalcommons.unl.edu/cgi/viewcontent.cgi?article= 1052&context=icwdmbirdcontrol (accessed 2 January 2019).

26 MacKinnon, B. (2004). Sharing the Skies, Chapter 13, Solutions: Lessons from Military Aviation Experience. https://www.tc.gc.ca/eng/civilaviation/publications/tTPp13549-chapter13-2139.htm (accessed 2 January 2019).

27 Krutkow, A., Pigłas, M., Smoliński, H., and Szymczak, J. (2017). Statistics and analysis of aircraft collisions with birds (bird strikes) in military aviation. *Aviation Advances & Maintenance* 40 (1): https://doi.org/10.1515/afit-2017-0001 (accessed 2 January 2019).

28 Dan, G. (2013). No. 100: Strike Incidents – Visualizing Data with ggplot2. http://genedan.com/no-100-strike-incidents-visualizing-data-with-ggplot2 (accessed 3 January 2019).

29 Copp, T. (2018). Military Times Crash Database. *Military Times* https://www.militarytimes.com/news/your-military/2018/04/06/military-times-aviation-database (accessed 3 January 2019).

30 Insinna, V. (2018). Aviation in Crisis: Wildlife strikes add to Air Force and Navy's mishap count. *Military Times* https://www.militarytimes.com/news/your-military/aviation-in-crisis/2018/04/14/wildlife-strikes-add-to-air-force-and-navys-mishap-count (accessed 3 January 2019).

31 Payson, R.P. and Vance, J.O. (1984). A Bird Strike Handbook for Base-Level Managers, M.Sc. Thesis, Air Force Institute of Technology, Air University. https://apps.dtic.mil/dtic/tr/fulltext/u2/a147928.pdf (accessed 3 January 2019).

32 Elsayed, A.F. (2016). *Fundamentals of Aircraft and Rocket Propulsion*. London: Springer.

33 Dolbeer, R.A., Wright, S.E., Weller, J. and Begier, M.J. (2014). Wildlife strikes to civil aircraft in the United States, 1990–2013. Report of the Associate Administrator of Airports Office of Airport Safety and Standards and Certification. Federal Aviation Administration National Wildlife Strike Database Serial Report 20. Washington, DC: Federal Aviation Administration.

34 Chow, D. 2014. Bird Strikes Problematic for Military Helicopters, Study Finds, *Live Science*. https://www.livescience.com/43833-military-helicopters-bird-threat.html (accessed 3 January 2019).

35 Frostenson, S. 2016. Planes killed 13,159 birds in 2014 – and one iguana. https://www.vox.com/2016/1/19/10789816/airplane-animal-kill (accessed 3 January 2019).

36 Pingstone, A. (2003). https://en.wikipedia.org/wiki/Boeing_737#/media/File:Lufthansa-1.jpg (accessed 3 January 2019).

37 https://upload.wikimedia.org/wikipedia/commons/6/61/Bmi_erj145_planform_arp.jpg (accessed 3 January 2019).

38 U.S. Coast Guard. https://en.wikipedia.org/wiki/Helicopter#/media/File:R-4_AC_HNS1_3_300.jpg (accessed 3 January 2019).

4

Fatal Bird Strike Accidents

Bird Strike in Aviation: Statistics, Analysis and Management, First Edition. Ahmed F. El-Sayed.
© 2019 John Wiley & Sons Ltd. Published 2019 by John Wiley & Sons Ltd.

4.1   Introduction

Bird strike on aircraft is an increasing economic and operational safety problem for the air transport industry worldwide; however, bird strike with military aircraft influences mission accomplishment as well as pilots' safety. Bird strike has also become a critical issue for both civilian and military helicopters.

The first bird strike was reported by Orville Wright in 1905, while the first fatal bird strike accident was encountered in 1912 [1].

The annual costs of bird strike in the civil aviation industry worldwide are at least US\$ 1.3 billion in direct damage. Annual costs in the USA alone are US\$ 951 million. Although the economic costs of bird strikes are extreme, the associated costs in human lives are beyond evaluation.

Fatal bird strike accidents are reviewed in many publications [2–8]. It has been estimated that there is only about one bird strike accident resulting in human death for every one billion (10^9) flying hours [7].

Concerning civilian aircraft and helicopters worldwide, during the period 1912 up to 2012, records identify that the number of fatal accidents had reached at least 55, killing 276 people and destroying 108 aircraft [7]. These accidents are categorized as follows:

- Airliners and executive jets: 16 fatal accidents, killing 189 people (which includes seven third parties on the ground) and destroying 44 aircraft;
- Airplanes 5700 kg and below: 32 fatal accidents, killing 69 people and destroying 56 aircraft;
- Helicopters: 7 fatal accidents, killing 18 people and destroying 8 helicopters.

Moreover, between 1990 and 2004, US airlines reported 31 incidents in which pilots had to dump fuel after striking birds on takeoff or during climbing. In each case, an average of 11 600 gal of jet fuel was dumped.

Looking at military aircraft in Europe and Israel [7] during the period 1912 up to 2012, records state that there were at least:

- 168 aircraft losses;
- 34 aircrew killed as well as three civilians on the ground.

4.2   Civil Aircraft

4.2.1   Introduction

First bird strike: on 7 September 1905, the first bird strike was encountered by Orville Wright. His aircraft hit a bird (probably a red-winged blackbird) as he flew over a cornfield near Dayton, Ohio.

First bird strike fatality: the first known fatal airplane crash involving a bird took place in 1912. A Wright Model EX plane (a single-seat exhibition model version of a 1911 Wright Model B) was flown by Calbraith Perry Rodgers (1879–1912). Rodgers was an American aviation pioneer who made the first flight across the USA (4321 miles), traveling between 17 September 1911 and 5 November 1911 and making 70 stops. On 3 April 1912, Rodgers was killed while testing a new engine. He flew into a flock of seagulls, hit them, and plunged into the surf. The engine broke loose and struck Rodgers in the back of his head, breaking his neck. The plane crashed into the ocean and Rodgers was the first to die.

4.2.2 Statistics of Annual Fatal Accidents Due to Bird Strike

Firstly, let us explore the whole spectrum of aircraft accidents caused by different causes and not only those due to a bird strike. The total number of fatal accidents for civilian aircraft in the period from 1 January 1960 to 31 December 2015 with a definitive known cause is 1104 accidents [9]. The planes involved include those with 10 or more passengers and one or more fatalities. (Small aircraft with passengers less than 10 are not included.) Whenever there were multiple causes, the most prominent cause was used.

The primary cause categories for such these accident are:

1. Pilot error, which includes improper procedure, descending below minima, excessive landing speed, missed runway, fuel starvation, navigation error, etc.;
2. Mechanical, which includes engine failure, equipment failure, structural failure, design flaw, maintenance error;
3. Weather, which includes severe turbulence, wind shear, poor visibility heavy rain, severe winds, icing, lightning strike, etc.;
4. Sabotage, which includes hijacking, being shot down, there being an explosive device aboard;
5. Other, which includes bird strike, ATC error, ground crew error, overloading, improperly loaded cargo, mid-air collision with other aircraft, pilot incapacitation, fuel contamination, obstruction on the runway, fire/smoke in flight.

Table 4.1 expresses the percentage of each possible cause for aircraft accidents during the five decades from the 1960s to 2000s. The table was compiled from the PlaneCrashInfo.com database [9].

Figure 4.1 illustrates the number of fatal accidents (civil aircraft with 19 or more passengers) in the period from 1950 to 2015. The highest number of accidents (41 accidents) was recorded in 1971 [9]. The number of accidents has continued to decrease since 1990 to reach its lowest value in 2015 (six accidents).

Figure 4.2 illustrates the number of fatal accidents in the period from 1958 to 2016 as documented by Airbus Industries [10]. The worst years, registering the largest number of accidents, were 1972 and 1988 (18 accidents), while the best years, with the smallest number of fatal accidents, were 1959 and 2015 (only one accident). These data confirm the positive effects of bird strike management and control. Though the number of flights has increased from 13 million in 1988 to reach 33 million in 2016 (×2.5), the number of fatal accidents has decreased, from 18 accidents to 5 in the same period (1988–2016) [10].

Figure 4.3 illustrates number of annual worldwide bird strike accidents in the period (1999–2008), as listed by Maragakis [11]. During the decade 1999–2008, 71 accidents in total occurred due to a bird strike. Of these only six led to fatal injuries.

Table 4.1 Aircraft accidents by causes.

Cause	1960s	1970s	1980s	1990s	2000a	Average
Pilot error (%)	60	55	54	60	60	58
Mechanical (%)	21	16	18	15	18	17
Weather (%)	6	5	6	6	7	6
Sabotage (%)	5	11	11	8	9	9
Other (%)	8	13	11	11	6	10

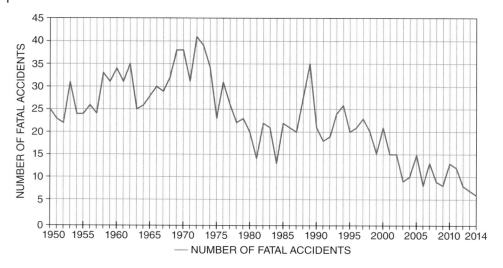

Figure 4.1 Number of fatal accidents in the period 1950–2015. *Source:* Reprinted with permission from planecrashinfo.com [9].

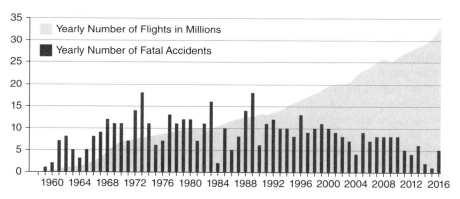

Figure 4.2 Yearly number of fatal accidents and the number of flights in the period from 1958 to 2016.

Figure 4.4 illustrates the yearly fatal accident rates per million of flights in the period from 1997 to 2017, as expressed by Airbus Industries [10]. The trend is for the rate to be continuously reducing. It has decreased from 0.5 in 1997 to 0.15 in 2016.

Some slightly different values for the yearly accident rate have been published by the Boeing Company [11] and these are shown in Table 4.2.

Figure 4.5 shows the number of fatalities in the period from 1950 to 2015 [9]. The worst year, with the largest number of fatalities, was 1972 (almost 2300) while the best year, having a minimum number of fatalities, was 2013 (nearly 200).

Bird strike belongs to the category "Others" in aircraft accidents (Table 4.1). "Others" represents on average some 10% of fatal accidents. Since this category includes eight different items, then we may assume the share due to bird strike to be 10–15%. Thus, bird strike contributes some 1–1.5% of the total number of accidents. From Figure 4.5,

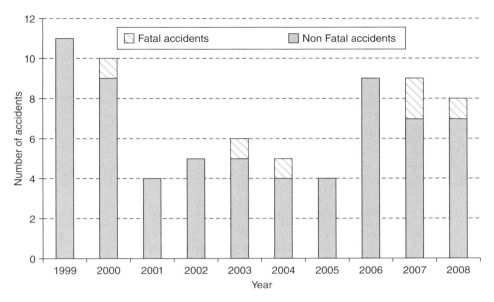

Figure 4.3 The annual worldwide fatal and non-fatal accidents due to bird strike in the period 1999–2008. *Source:* Courtesy IBSC [11].

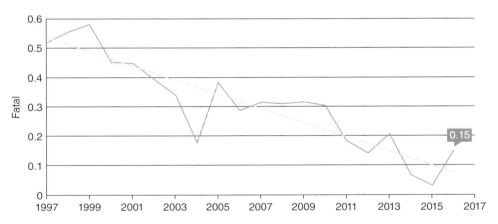

Figure 4.4 Yearly fatal accident rates per million of flights in the period from 1997 to 2017 [6]. *Source:* Correction, Courtesy IBSC [11].

Table 4.2 Yearly fatal accident rates per million of flights.

	1997	1999	2001	2003	2005	2007	2009	2011	2013	2015
USA and Canadian operators	0.5	0.35	0.4	0.15	0.15	0.15	0.55	0.15	0.3	0.0
Rest of the world	0.75	1.0	0.9	0.6	0.75	0.7	0.6	0.2	0.15	0.15

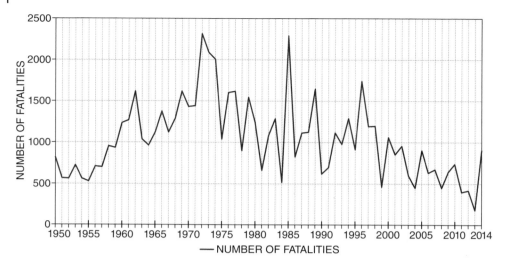

Figure 4.5 Number of fatalities in the period from 1950 to 2015. *Source:* Reprinted with permission from planecrashinfo.com [9].

the minimum and maximum numbers of fatalities are 200 and 2300, respectively , so the expected numbers of fatalities should have a minimum value of 2 and a maximum of 35. However, based on Airbus Industries data, these figures will be lower (Figure 4.2).

4.2.3 Statistics of Critical Flight Phases

Airbus Industries divide any aircraft trip into 11 flight phases [10]. These are: parking, taxi; takeoff run, aborted takeoff, initial climb, climb to cruise, cruise, initial descent, approach, go-around, and landing. The fatal and non-fatal accidents encountered in these different flight phases in the period 1997–2016 are plotted in Figure 4.6.

Boeing has a slight different identification of flight phases (as compared to Airbus Industries, above) and these are plotted in Figure 4.7. In 2017, the Boeing Company published statistics for accidents that occurred in the period 1959–2016 [12]. Table 4.3 lists the percentage of fatal accidents as well as the percentage of onboard fatalities associated with different flight phases, but Table 4.3 only covers the period 2007–2016.

As outlined by Maragakis [11], among the 71 accidents due to bird strike in the decade 1999–2008, only six led to fatal injuries (Figure 4.8). Based on the flight phase, the 71 accidents may be categorized as follows:

- 48% during the takeoff phase
- 30% during approach
- 6% during landing
- 15% en-route phase

Here the "en-route" phase includes the climb and descent phases.

Thus, in total 84% of bird strike accidents occurred during the takeoff, approach, and landing phases. Bird strike during the takeoff phase (acceleration and lift-off) is more dangerous than other flight phases as the aircraft will be susceptible to partial or total

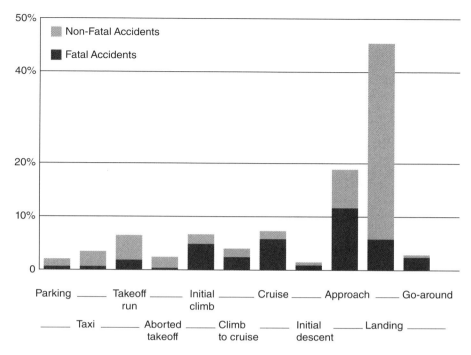

Figure 4.6 Accidents by flight phase as a percentage of all accidents, 1997–2016.

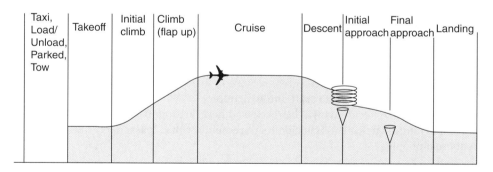

Figure 4.7 Different phases of flight.

loss of control. Such a loss of control accompanied by the high speed of aircraft and its proximity to the ground will contribute significantly to damage to the aircraft and passengers/crew injuries.

4.3 Fatal Accidents of Civil Aircraft

A detailed database for fatal accidents is described by Thorpe [5]. The foundation of this database is attributed to the lifetime of work by John Thorpe, John Richardson, and

Table 4.3 Percentages of fatal accidents (2007–2016) [12].

	Phases of flight								
	Taxi, load/ unload, parked, tow	Takeoff	Initial climb	Climb (flaps up)	Cruise	Descent	Initial approach	Final approach	Landing
Fatal accidents (%)[a]	10	6	6	6	11	3	8	24	24
Onboard fatalities (%)[a]	0	6	1	7	22	3	16	26	20

a)   The percentages may not sum to 100% due to numerical rounding [12].

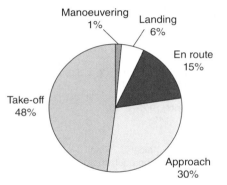

Manoeuvering 1%
Landing 6%
En route 15%
Take-off 48%
Approach 30%

Figure 4.8 Phase of flight during which the bird strike occurred and led to an accident, worldwide (1999–2008).

Tim West. They have performed a great service in analyzing numerous wildlife strikes for both civilian and military aircraft and helicopters.

Most of these fatal accidents will be discussed here, arranged in chronological order.

It is important in such a list of accidents to remember that "passengers and crew" are actually real people.

10 February 1939

An Arado aircraft powered by a piston engine was flying from Tamil Nadu, Madras province. During climb at an altitude less than 100 ft above ground level (AGL) the plane struck an unknown bird. The aircraft was destroyed and the two occupants were killed.

1955

A Cessna airplane powered by a piston engine was flying from Aberdare Mountains, Kenya (date is unknown). A single old-world vulture hit a wing tip, leading to the ailerons jamming. The aircraft crashed and the pilot died.

Figure 4.9 Lockheed Electra. *Source:* Courtesy IBSC [5].

4 October 1960

Eastern Airlines Flight 375, a four-engine Lockheed Electra turboprop (Figure 4.9) took off from Boston Airport, Massachusetts, half an hour before sunset with 72 people on board [3].

Soon after it began its takeoff (27 seconds) and only 7 seconds into the flight, it struck a flock of European starlings (*Sturnus vulgaris*, 80 g) estimated at 10 000–20 000 birds. At least four birds were ingested into engine number 1 (on the left wing farthest from the fuselage). This engine shut down, giving off a puff of gray smoke. The plane continued flying, but within seconds, about six birds were ingested into engine 2 (on the left wing closest to the fuselage) which flamed out, shut down, and re-lit. Then a smaller number of birds were ingested into engine number 4 (on the right wing farthest from the fuselage), which also lost power. The plane crashed into Boston Harbor (Figure 4.10), with 62 of 72 on board being killed (59 passengers and three crew), and nine passengers being seriously injured. Following this accident, the Federal Aviation Administration (FAA) initiated action to develop minimum bird ingestion standards for turbine-powered engines [3].

15 July 1962

An Indian Airlines Douglas DC3 (powered by two P&W R1830 piston engines) was on a freight flight from Kabul to Amritsar. When in the cruise phase the crew spotted a

Figure 4.10 Lockheed Electra Tail section lifted out of the harbor. *Source:* Courtesy IBSC [5].

vulture (weight up to 10 kg) above and to one side of them. The vulture smashed through the windscreen, killing the co-pilot.

23 November 1962

A United Airlines flight using a Vickers Viscount (Figure 4.11) powered by four Rolls Royce Dart turboprop engines was cruising at an altitude of 6000 ft en-route from Newark Airport (New York) to Washington at night. The aircraft penetrated a flock of whistling swans (*Cygnus columbianus*, 6 kg). Two swans were struck. One penetrated the leading edge of the left horizontal stabilizer and exited from the rear surface, damaging the elevator, weakening the structure, and causing the stabilizer to detach. The aircraft became uncontrollable and struck the ground in a nose-low inverted attitude. The crash (Figure 4.11) resulted in 17 fatalities [7].

28 July 1968

A Falcon 20 aircraft, powered by two GE CF700 engines, was taking off from Burke Lakefront Airport, Cleveland, USA, when many gulls (280 g to 1.7 kg.) were ingested into both engines causing severe damage [7]. The aircraft collided with a fence and crashed landed in the Lake Erie, where the three crew were rescued by a pleasure boat. A total of 315 birds were found dead on the runway. Engine number 1 was 20% filled with debris and the compressor of engine number 2 by 17% filled with the remains of birds.

23 July 1969

An Air Djibouti Douglas DC3 aircraft was operating a freight flight from Tadfours to Djibouti. While flying at 300 ft (91 m) a flock of cranes (up to 6 kg) struck the propellers

Figure 4.11 Vickers Viscount accident. *Source:* Courtesy IBSC [7].

mounted to the radial piston engines and the debris blocked both carburetor intakes. The aircraft was ditched in the sea 9 nautical miles from Khar Ambadu and the four crew members were rescued by a passing boat [7].

26 February 1973

A Lear 24 aircraft operated as a corporate flight by Machinery Buyers Corp and powered by two GE CJ610 engines collided with a flock of cowbirds (44 g) just after takeoff from De Kalb, Chamblee, Georgia, USA. The left engine had 14 strikes and the right at least 5. As a result, both engines lost power. The aircraft crashed into buildings and burned. There were seven fatalities and one person on the ground was injured [13]. Post-crash investigations revealed the presence of bird residue and feather on the airplane's windshield and center post. Remains of 15 cowbirds were found within 150 ft of the departure end of the runway. Both engines showed damage to the compressor rotor assemblies and obstruction of some 75% of the cooling air ports of the first-stage turbine nozzles.

4 December 1973

A BAC one-eleven aircraft operated by Austral Lineas Aereas suffered problems shortly after lift-off from Bahia Blanca, Argentina. While retracting the landing gear there was a loss of power and severe vibration from the left engine (#1 engine) and the aircraft lost height [14]. The pilot had seen a large bird on the left side of the aircraft. He attempted to land back on the remaining 950 m of runway and was slowed by arrester cables used for the operation of navy A4Q fighters. The cables broke, damaging the aircraft and puncturing a fuel tank, which caused a fuel leak. The fuel then ignited by friction sparks. The aircraft was damaged beyond economic repair but there were no fatalities.

12 December 1973

A Falcon 20, operated by Fred Olsen Flyveselskap, and powered by two GE CF700-2C engines, was on a charter flight from Norwich, Norfolk, UK, to Gothenburg, Sweden, at 15.37 hours [7]. As it became airborne about halfway down the runway, the pilot avoided two flocks of birds but between 100 and 200 ft collided with a third flock extending from the ground to well above the aircraft. There were multiple strikes, and both engines failed. The landing gear was still down. Avoiding trees, the pilot force-landed in a field about 1000 m off the runway end. All three landing gear legs were torn off, and the plane came to rest on its belly (Figure 4.12). The two pilots and the cabin attendant suffered cuts and bruises, but the passengers were uninjured.

24 June 1974

An Ilyushin Il-18D aircraft operated by Aeroflot was in its initial climb from Tashkent, Uzbekistan. The plane, powered by four Ivchenko Al 20M turboprop engines

Figure 4.12 Falcon 20 accident. *Source:* Courtesy IBSC [7].

Figure 4.13 Ilyushin Il-18D aircraft [60]. *Source:* Rossiya Ilyushin Il-18 by Sergey Riabsev – http://www2.airliners.net/photo/Russia-State-Transport/Ilyushin-Il-18/1206404/L License GFDL 1.2.

(Figure 4.13), ingested a bird into engine #4 and lost power. The aircraft crashed and was destroyed. One passenger out of 114 occupants died because of the accident.

12 November 1975

An Overseas National Airways (ONA) McDonnell Douglas DC-10-30CF aircraft powered by GE CF6 engines was in a ferry flight from John F. Kennedy International Airport (New York) to Jeddah via Frankfurt, having 114 passengers (all ONA employees). During its takeoff roll the plane struck a flock of seagulls (about 100, including herring gulls and great black-backed gulls), causing the crew to perform a rejected takeoff. The fan blades of engine #3 were damaged, and the casing of the high-pressure compressor was separated [15]. Several tires and wheels disintegrated and the landing gear collapsed. A resulting fuel leak led to an extensive ground fire which destroyed the airplane. Being highly trained in emergency evacuation, all personnel on board were evacuated safely. Of the 128 passengers and 11 crew members on board, only two people were seriously injured and 30 were slightly injured (Figure 4.14).

20 November 1975

A British Aerospace BAe 125 aircraft, powered by two RR Viper 601 engines and operated by Hawker Siddeley Aviation Limited, departed from Dunsfold, Surrey, UK, with two pilots and seven passengers on board [16, 17]. The aircraft was at about 50–100 ft, just after becoming airborne, when it encountered a large flock of lapwings (*Vanellus*, 215 g). Both engines lost power due to surge conditions. The pilot attempted to land back on the runway. However, the aircraft over-ran the runway end, crossed a road, and hit a deep ditch. The collision ruptured and initiated the detachment of the entire undercarriage, before the plane was destroyed by fire (Figure 4.15). The nine occupants

Figure 4.14 DC-10-30CF accident. *Source:* Courtesy NTSB [15].

Figure 4.15 BAe 125 aircraft accident. *Source:* Courtesy IBSC [7].

safely evacuated; the two pilots were slightly injured while the seven passengers were unhurt. However, during the overrun when the aircraft was crossing the road, it struck and destroyed a passing private Ford Cortina motor car, killing all six of the occupants of the car.

4 April 1978

A SN Boeing 737-229C, powered by two P&W JT8D engines and operated by Sabena Airline, was on a training flight with an instructor and two co-pilot students. The plane departed from Charleroi-Gosselies Airport, Belgium. Both the students were going to practice ILS approaches to runway 25 followed by a touch-and-go. During one of the touch-and-go procedures, a flock of birds (ring doves) were observed crossing the runway [18]. Several birds were ingested as the airplane was rotating. The instructor took over control and attempted to continue the takeoff. The airplane failed to respond to his control inputs and seemed to decelerate. He then decided to abort the takeoff. There was insufficient runway length available, so the plane overran, struck localizer antennas, and skidded. The right main landing gear collapsed and the #2 engine was torn off in the slide. The aircraft came to rest 300 m past the runway end and was destroyed by fire (Figure 4.16). The three crew escaped. Just one bird had written off a US$ 20 million aircraft.

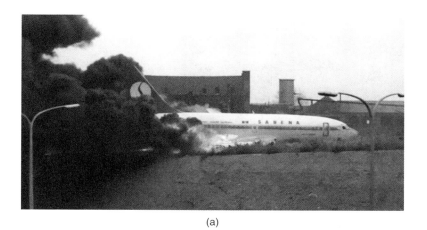

(a)

(b)

Figure 4.16 SN Boeing 737-229C aircraft accident. (a) Aircraft on fire. (b) Complete destruction. *Source:* Courtesy IBSC [7].

Figure 4.17 Convair CV-580 aircraft accident. *Source:* Courtesy IBSC [7].

25 July 1978

A North Central Airlines aircraft Convair CV-580 aircraft, powered by two Allison 501 turboprop engines, was taking off from Kalamazoo, Michigan, USA at 07.02. A sparrowhawk, weighing about 120 g, was ingested into the left engine just as the aircraft passed V_1 [19]. The left propeller feathered itself automatically during lift-off. This caused the aircraft to turn left, due to asymmetric thrust, and crash in a cornfield Figure 4.17. Moreover, the compressor experienced a transient stall. One crew member of the crew of three and two passengers out of 40 were seriously injured. The probable cause of the accident was the failure of the captain to follow the prescribed engine-out procedures, which allowed the aircraft to decelerate into a flight regime from which he could not recover. Contributing to the accident were inadequate cockpit coordination and discipline.

6 December 1982

As a Lear 35A aircraft operated by Transair SA and powered by two Garrett TFE731 turbofans was taking off on a wet runway from Paris Le Bourget, France, it struck a flock of black-headed gulls [20]. Takeoff was aborted when the aircraft had already accelerated through V_1. The aircraft overran the runway by 56 m and collided with a localizer antenna, which penetrated the cockpit and injured the co-pilot. The brake chute had failed and the emergency brake source was not used. Neither the two crew nor the two

Figure 4.18 Lear 35A aircraft accident. *Source:* Courtesy IBSC [7].

passengers were killed. However, the airplane was written off (damaged beyond repair) (Figure 4.18).

15 September 1988

Ethiopian Airlines flight 604, a Boeing 737-200, took off in a scheduled domestic flight (Addis Ababa to Bahir Dar to Asmara). At 5800 ft above mean sea level (AMSL), and at a speed between V_1 and V_R, the aircraft encountered a flock of speckled pigeons (*Columba guinea*, 320 g). Some 10–16 birds were ingested in each engine. One engine lost thrust almost immediately and the second lost thrust during the emergency return to the airport, leading the crew to execute a wheels-up landing. During such a gear-up landing, the aircraft caught fire (Figure 4.19). All six crew members survived, but 35 of the 98 passengers were killed and 21 were injured. Twelve of those injured were in a serious condition [21].

Wreckage of Boeing 737-200   **Speckled pigeon**

Figure 4.19 Wreckage of Boeing 737–200 and a speckled pigeon. *Source:* Courtesy IBSC [7].

Figure 4.20 Antonov 124 [61]. *Source:* An-124-100 of Maximus Air Cargo at Brno Airport (2010) by Marek Vanzura – http://www.airspotter.eu/photos/2010/2805/ur-zydb.jpg License GFDL 1.2.

13 October 1992

An Antonov freighter, the second An-124 prototype (Figure 4.20), was operating on a test flight after departing near to Ulyanovsk (Kiev, Russia). At about 19 700 ft during a high-speed descent (330 km/h), a bird (1.85 kg) was struck, holing the nose. This allowed the area between the nose and the front bulkhead to become pressurized by the ram-air, causing failure of the upward opening freight door. Control was lost and the aircraft crashed. One of the crew managed to eject, but the eight others were killed when the plane crashed in a forest, as it was outside the normal flight envelope.

20 August 1993

An Antonov An-12 aircraft (Figure 4.21), powered by four Ivchenko AL20 turboprop engines, was on a freight flight and was climbing at about 150 ft when engines #2 and #4

Figure 4.21 An-12 [62]. *Source:* Shaanxi Y-8 of the Myanmar Air Force. M Radzi Desa – http://www
.airliners.net/photo/Myanmar---Air/Shaanxi-Y-8/1643688/L License GFDL 1.2.

failed. It is believed the engine failures resulted from multiple ingestions of birds in the
vicinity of the runway [7]. The crew attempted to return but had to force land beyond
the end of the runway. The plane touched down with the landing gear retracted and slid
for about 460 m before it caught fire and was destroyed, resulting in seven fatalities.

3 June 1995

An Air France Concorde (Figure 4.22) ingested one or two Canada geese into the engine
#3 when landing at John F. Kennedy International Airport at only 10 ft AGL. The engine

Figure 4.22 Air France Concorde [63]. *Source:* Air France Concorde at CDG Airport in 2003. Alexander
Jonsson – http://www.airliners.net/photo/Air-France/Aerospatiale-British-Aerospace-Concorde/
0432634/L License GFDL 1.2.

Figure 4.23 Merlin III aircraft [64]. *Source:* https://en.wikipedia.org/wiki/Swearingen_Merlin#/media/ File:Fairchild_Swearingen_Merlin.jpg.

failed and suffered an uncontained failure. Shrapnel from the engine #3 destroyed engine #4 and cut several hydraulic lines and control cables. The pilot was able to land the plane safely, but the runway was closed for several hours. Damage to the Concorde plane was estimated at over US$ 7 million. The French Aviation Authority sued the Port Authority of New York and New Jersey and eventually settled out of court for US$ 5.3 million.

4 April 1996

A Merlin III aircraft (Figure 4.23), powered by two Garrett TPE 331 turbofan engines, struck several large birds while landing in Ushuaia, Argentina. One bird broke the windshield and others struck the left engine. Control was lost, and the aircraft ran off the side of the runway and was damaged beyond repair.

27 July 1998

An Antonov An-12 aircraft, powered by two Ivchenko AI-20 turboprop engines, was taking off from St Petersburg, Russia, at 03.42 with 13 tons of freight, seven crew, and two passengers. Immediately after lift-off, one engine suffered bird ingestion (possibly crows and gulls). The pilot lost control. The aircraft descended from about 600 ft onto the runway and caught fire. All occupants safely escaped but one suffered severe burns.

19 April 2000

A Central African Airlines Antonov An-8 aircraft (Figure 4.24) powered by two Ivchenko AI-20 turboprop engines, was on a flight from Pepa, Zaire, to Kigali, Rwanda. Shortly after takeoff the plane suffered bird ingestion and crashed while attempting to return to the airstrip. All 24 people on board were killed.

1 June 2003

A Eurojet Italia Lear 45 aircraft, powered by two TFE 731 Turbofan engines, was taking off from Milan Linate for Genoa (Italy) with two crew on board. A flock of feral pigeons

Figure 4.24 Antonov An-8 aircraft [65]. *Source:* Aeroflot Antonov An-8, Charles Osta – http://www.airliners.net/photo/Aeroflot/Antonov-An-8/1975766/L License GFDL.

struck the aircraft and were ingested into at least one engine. The Learjet lost control and crashed into a warehouse; fortunately the warehouse was unoccupied (it was a Sunday) but it was destroyed by fire. The two crew members were killed.

8 July 2003

A Cessna 172S single-engine aircraft hit a bird (likely a vulture) with the left wing while at 800 ft AGL. The plane was approximately 2 miles west of Aero Country Airport (T31), near McKinney, Texas. The pilot attempted to make an emergency landing in a field but lost control, impacted the ground, and crashed. The flight instructor and student pilot were killed (Figure 4.25) [22].

Figure 4.25 Cessna 172 accident. *Source:* Courtesy USDA-APHIS [22].

24 October 2004

A Boeing 767 struck a flock of birds during its takeoff run when departing Chicago O'Hare Airport. As a result, the compressor stalled, and the engine flamed out. The pilot dumped approximately 11 000 gal of fuel.

28 March 2006

An Antonov 12 cargo airplane had departed from Payam Airport, Iran, flying to Sharjah (SJH). The aircraft encountered a flock of birds immediately after takeoff, causing engines #1, #3, and #4 to fail. The pilot tried to return to Payam, but an emergency landing had to be made about 3 miles from the airport. The aircraft broke up and caught fire.

23 November 2008

United Airlines Flight 297, a Vickers Viscount 745D, was flying at 6000 ft, en route from Newark to Washington when it encountered a flock of whistling swans. At least two of the swans struck the aircraft. One swan penetrated the leading edge of the left horizontal stabilizer, damaged the elevator, and rendered the plane uncontrollable. The aircraft struck the ground in a nose-low inverted attitude near Ellicott City, Maryland, following this bird strike. All 17 people (4 crew and 13 passengers) on board died.

25 May 2009

A Boeing 747F-200 operated by Kalitta Air LLC was on a cargo flight from New York (JFK) to Bahrain, with a technical stop at Brussels (Belgium), with 73 tons of cargo. When the Boeing 747 was taking off from Brussels Airport, the right engine experienced a kestrel strike. This engine experienced a momentary loss of power accompanied by a loud bang (heard by the crew and external witnesses) and flames seen from the control tower [6]. The takeoff was abandoned. All four engines were brought back to idle, braking action was initiated, but the thrust reversers were not deployed. The aircraft came to a stop 300 m beyond the end of the runway. The aircraft was severely damaged and broke into three parts (Figure 4.26). The four crew members and a passenger were safely evacuated and suffered only minor injuries.

15 January 2009 – The "Miracle on the Hudson"

This outcome of this accident is called a "miracle" because a brilliant pilot (Captain Chesley B. "Sully" Sullenberger), as well as skilled teamwork, saved all the people on board. An Airbus A320-214 aircraft lost power from all engines due to a bird strike. The pilot glided a powerless giant airliner to a safe landing on a Hudson River, "landing" 6 minutes after takeoff. Captain Sullenberger saved the lives of 155 passengers and crew. The US Airways Flight #1549 Airbus A320-214 took off at 15:26 from New York La Guardia's airport for Charlotte, Carolina [7]. The first officer was the handling pilot. As the plane reached an altitude of 3200 ft, the crew encountered a flock of Canada geese (*Branta Canadensis*, 3.6 kg). Impacts were felt, both engines began to lose power, and there was a burning smell.

Figure 4.26 Boeing 747-200 accident. *Source:* Courtesy IBSC [6].

The captain took over control of the flight while the first officer attempted to relight the engines. ATC was informed that they had lost thrust in both engines and were turning back toward LaGuardia. It quickly became evident that they were not able to reach either LaGuardia airport or Teterboro airport (New Jersey). The pilots stated their intention of going for the Hudson River [23].

They descended over the George Washington Bridge and ditched opposite mid-town Manhattan (Figure 4.27a). The last radar return received was at 300 ft and 153 knot. The ditching looked like a "hard landing" with "one impact, no bounce, then a gradual deceleration". The passengers were evacuated through the four over-wing window exits and into an inflatable slide/rafts. All 155 people aboard were rescued by nearby Coast Guard, commuter, and tour vessels. Only four passengers and one flight attendant were seriously injured.

The aircraft was retrieved from the river. Both DNA and feather analysis confirmed that both engines had ingested Canada geese. This accident is classified as "fatal", due to the severe damage to the aircraft (Figure 4.27b), even though there were no causalities.

This "Miracle on the Hudson" has acted as an alarm call, awakening all aviation authorities and personnel to the danger of catastrophic airliner crashes caused by a bird strike.

30 September 2009

Germania airlines B737 flight from Dusseldorf to Kosovo had 80 people on board. As it lifted off at 200 mph, a flock of more than 200 starlings (which looked like a scene from the Hitchcock movie *The Birds*) were sucked into the right engine. Others birds dented the fuselage but did not pierce it [24].

6 June 2010

A Royal Air Maroc Boeing 737-400 was forming flight AT-685 from Amsterdam (Netherlands) to Nador (Morocco) with 156 passengers and six crew. Upon departing

(a)

(b)

Figure 4.27 Airbus A320-214 ditched in Hudson River. (a) Evacuation of passengers. (b) Damage to aircraft. *Source:* Courtesy NTSB [57].

Schiphol Airport's runway 18L, the plane collided with a flock of Canada geese. The right-hand engine (CFM56) was seen to be on fire during initial climb, prompting the crew to shut the engine down and activate the engine's fire suppression system. The fuselage received substantial damage (Figure 4.28). The airplane returned to the airport for a safe landing after 17 minutes. Six dead birds found on the runway [25, 59].

Post-impact examination revealed dents in the underside of the fuselage near the nose of the aircraft, a dent in the leading edge of the vertical fin, and that the left-hand engine had the following damage:

- three fan blades fractured at about mid-span damaging all the rest of the fan blades;
- damage to the low and high-pressure compressors

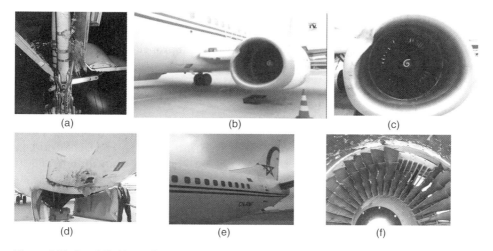

Figure 4.28 Royal Air Maroc (Boeing 737-400) [59]. (A) Landing gear damage, (B) engine inlet impact, (C) inlet cowl damage, (D) inlet cowl damage, (E) dent in the leading edge of the vertical fin, (F) fractured fan blades.

- damage to the combustion chamber;
- damage to the high-pressure turbine stator and rotor blades;
- the first to fourth low-pressure turbine stages were all damaged;
- damage to the low-pressure turbine outlet guide vanes.

11 April 2010

A Piper PA24 Comanche plane, flying at 1200 ft AGL in Ohio, struck a bald eagle. Both the windshield and engine cowl were damaged (Figure 4.29).

30 June 2011

A Piper PA31-350 Navajo Aircraft struck a large bird during landing, just 5–10 ft before touchdown. The bird broke and penetrated the front center windshield. The

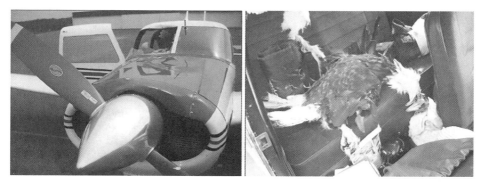

Figure 4.29 PA24 Comanche aircraft accident. *Source:* Courtesy IBSC [7].

pilot reacted by leaning to the right. The aircraft wing contacted the runway, causing the aircraft to nose into the ground. The aircraft suffered severe damage to the left wing and the left landing gear collapsed. Moreover, both propellers were bent and the aircraft was damaged beyond economic repair. The bird species is unknown.

28 September 2012

A Sita Air Dornier Do 228 was performing domestic Flight 601 from Tribhuvan International Airport in Kathmandu (the capital of Nepal) to Tenzing-Hillary Airport in Lukla with 16 passengers and three crew. At 06:17 local time and at an altitude of 50 ft (15 m), a black kite collided with the right-hand engine. Some parts of the engine separated, impacted the vertical tail, and disabled the rudder. This made the aircraft uncontrollable. The pilot requested to fly back to the airport, but the aircraft began to sway and perform unusual maneuvers. The aircraft next descended near the Manohara River, nose-dived, and crashed on the banks of the Manohara River. The aircraft burst into flames after coming to rest. The front part of the fuselage was completely destroyed (Figures 4.30 and 4.31).

Several people survived the crash and were screaming for help inside the burning wreckage. However, due to delays to the emergency services, most parts of the plane were destroyed and all 19 people on board perished in the accident.

Figure 4.30 Wreckage of Dornier Do 228, Flight 601 immediately after impact. *Source:* Courtesy IBSC [7].

Figure 4.31 Remains of empennage of Dornier Do 228. *Source:* Courtesy IBSC [7].

13 March 2014

The #2 (right) PWC JT15D-5 engine of Guardian Pharmacy LLC Beechjet 400A aircraft sustained a herring gull bird strike during its takeoff climb from Rochester International Airport (Rochester, NY) at 10:20 eastern daylight time. The flight crew declared an emergency and returned to the airport, where they performed an uneventful landing. The impact from the bird strike resulted in contact between the fan and inlet case that separated 11 of the 19 fan blades (Figure 4.32) and subsequently led to the failure of the No. 1 bearing housing. A midshaft fracture of the low-pressure turbine shaft and inlet cowl was observed. The compressor cowl had multiple holes, and the right wing had an impact mark forward of the engine fan case. Several weeks after the event, a blade fragment was found lodged in the roof of a building below the flightpath [26].

7 January 2015

A Delta Air Lines Boeing 757-200 aircraft was climbing from Portland International Airport (Oregon) and was at 972 ft AGL when five northern pintail birds impacted the aircraft. At least one was ingested into engine #1, damaging fan blades and cowling, leading to bad vibrations. The aircraft was out of service for 11 days. The cost of the repairs was US$ 5 million, while other costs were US$ 452 320.

19 May 2016

A Robin DR 400 aircraft had taken off from Coimbra, Portugal, and was heading for Dax, Spain, when it struck a vulture and crashed, killing three people.

Figure 4.32 Damage to the inlet cowl of the #2 engine. *Source:* Courtesy USDA-APHIS [28].

29 July 2017

An Antonov 74TK-100 cargo plane was taking off from São Tomé Island Airport (TMS) when a bird struck engine #2. The flight crew aborted the takeoff, but the aircraft failed to stop on the remaining runway. It went down an embankment and came to rest next to the perimeter fence. This resulted in breaking the forward fuselage just in front of the wings, stowing of the thrust reverser on engine #1 and deploying the thrust reverser of engine #2.

26 September 2017

A Badr Airline Boeing 737 was on a domestic flight (J4 341) from Ad Damazeen to Khartoum International Airport (both in Sudan) when it encountered multiple bird strikes 45 minutes into the flight [27]. The radome was severely damaged. The crew managed to fly the plane safely to its destination.

4.4 Statistics for Civil Aircraft Accidents

4.4.1 Statistics for Critical Damaged Parts of Aircraft

The aircraft parts struck in fatal accidents discussed in Section 4.3 are displayed in Figure 4.33. This data is for transport airplanes and executive jets for the 100 year period 1912–2012 [7]. Most accidents were due to engine damage (76% of the accidents) followed by windshields, which were only 7%.

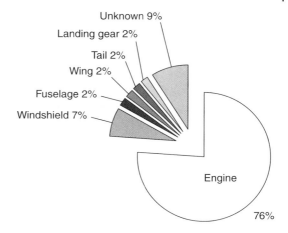

Figure 4.33 Part struck in fatal accidents for transport airplanes and executive jets in 100 years (1912–2012). *Source:* Courtesy IBSC [7].

Since the impact force is proportional to the square of the collision speed, then a small increase in speed results in a big increase in the impact force. Several things may be noted here:

- Pilots who fly fast at low altitudes or in a bird-rich environment should wear head protection with a visor or goggles. Modern aircraft have a considerably higher cruising speed than before.
- Single-engine training aircraft such as the Cessna 152 and Piper PA28 which fly at a modest speed (up to 80 knots) will enable birds to get out of their way.

Looking at smaller aircraft, less than 5700 kg, Figure 4.34 illustrates the plane part struck in fatal accidents for this group of airplanes for the 100 year period (1912–2012).

As shown in Figure 4.35, the most critical part is the windshield, which is struck in 56% of accidents. The engine encounters 13%, while the wing encounters 12% of accidents. Flight controls are subjected to 10% of accidents while tail experiences 7% of accidents. The lowest accident percentage, at 2%, that is encountered by the propeller.

In statistics for the decade 1999–2008, which combined all types of aircraft [11], the location where damage was sustained is shown in Figure 4.35. The engine is subjected to the highest number of strikes as illustrated for transport airplanes (Figure 4.33).

Figure 4.34 Part struck in fatal accidents for small airplanes for 100 years (1912–2012). *Source:* Courtesy IBSC [7].

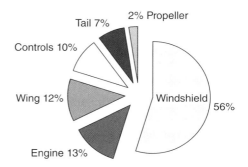

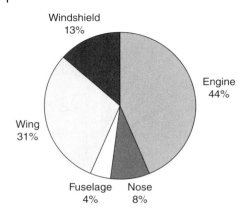

Figure 4.35 Part struck in fatal accidents for airplanes in the decade 1999–2008.

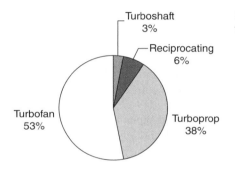

Figure 4.36 Percentages of fatal strikes with different types of engines in the decade 1999–2008.

4.4.2 Statistics for Strikes with Different Types of Engines

As analyzed by Maragakis [11] for the decade 1999–2008, more than half of the aircraft which sustained engine damage were powered by turbofan engines (Figure 4.36). Aircraft powered by turboprop engines represent 38% of the total accidents. Both aircraft powered by turboshaft engines and piston engines are subjected to a lower number of strikes since they have small engine inlets as well as being protected by the rotor or the propeller.

4.4.3 Effects of the Wildlife Strike on the Flight

In the period 2008–2015, 12 227 cases of bird strike were recorded [27]. Out of those, 2501 (which is 20% of total strikes) had a clear influence on the flight, while the rest had either no effect or no clear indication. These 2501 cases may be categorized as follows:

- 1230 cases of precautionary landing (highest effect)
- 513 cases of aborted takeoff
- 63 cases of the engine(s) being shut down.
- 211 cases of delayed flights
- 137 aircraft forced to return
- 54 cases of declaring a technical emergency

Figure 4.37 Bird species responsible for fatal accidents of transport airplanes and executive jets in 100 years. *Source:* Courtesy IBSC [7].

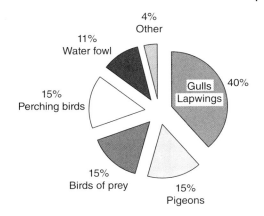

4%
Other

11%
Water fowl

15%
Perching birds

Gulls
Lapwings   40%

15%
Birds of prey

15%
Pigeons

4.4.4 Dangerous Birds

The study by Thorpe [7], identifies the most dangerous birds impacting large transport airplanes and executive jets. As illustrated in Figure 4.37, gulls and lapwings are the major cause of accidents (40%), followed by birds of prey, pigeons, and perching birds (each responsible for 15% of accidents). Waterfowl are responsible for 11% of accidents.

Figure 4.38 illustrates the birds influencing the safety of small aircraft (less than 5700 kg). Birds of prey caused 47% of the accidents. The next group comprises geese and ducks, causing 21% of accidents, followed by gulls (13%). The pelican/cormorant and "other water birds" are each responsible for 8% of accidents. The least accident percentage (3%) is due to the jackdaw.

Figure 4.39 displays the statistics from the European Aviation Safety Agency (EASA) covering the period 1999–2008 [11]. It shows that most of the bird strikes during these 10 years were due to flocks of large birds (45%), followed by strikes from single large birds (31%). Moreover, most of these strikes were from geese, ducks, cormorants, and hawks.

Figure 4.38 Bird species responsible for fatal accidents of small aircraft in 100 years. *Source:* Courtesy IBSC [7].

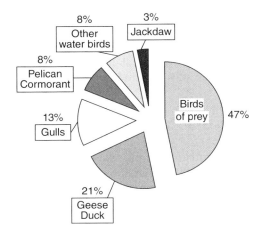

8%
Other
water birds

3%
Jackdaw

8%
Pelican
Cormorant

13%
Gulls

Birds
of prey   47%

21%
Geese
Duck

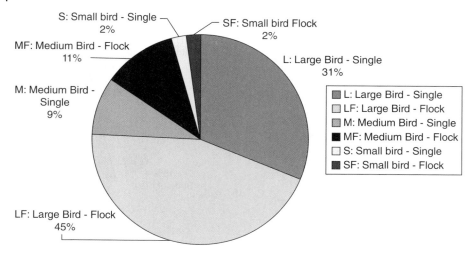

Figure 4.39 Types of bird species involved in bird strike accidents, worldwide (1998–2008).

4.5 Statistics for Bird Strike Incidents/Accidents in the USA (1990–2015)

Table 4.4 lists some statistics for bird strikes given in bird species, and the associated fatalities or seriously injured personnel in the USA for 25 years [28]. There were 229 fatal accidents, totaling some 400 human injuries.

Table 4.5 lists the minimum projected annual losses in aircraft downtime (hours) and repair and other costs from wildlife strikes with civil aircraft in the USA in the period 1990–2015. Total losses reached US$ 191 million [28].

Table 4.6 lists 164 444 strike reports involving birds from 1990 to 2015. From these reports, 13 558 (8%) indicated damage to the aircraft. When classified by level of damage, 7230 (4%) indicated the aircraft suffered minor damage, 3460 (2%) indicated the aircraft suffered substantial damage, 2831 (2%) reported an uncertain level of damage, and 37 reports (less than 1%) indicated the aircraft was destroyed because of the bird strike [28].

Finally, as identified by Kitowski [29], gulls and diurnal raptors present the greatest danger for both military and civil air traffic in Europe. Less dangerous birds turn out to be swallows, swifts, pigeons, European starlings and northern lapwings. Storks, herons, and vultures are dangerous locally. The most tragic accidents are caused by gulls, European starlings, and northern lapwings.

4.6 Statistics for Russian Accidents (1988–1990)

Some statistics for accidents due to bird strike with aircraft in Russia during the period 1988 to 1990 are listed by Sergey [30]. Before 1988 every strike was investigated by a special temporary board which included the maximum available information about the strike. Reports were transmitted to central branch authorities. After 1988, all bird strike events have been registered on a special form and sent to central authorities.

Table 4.4 Number of strikes to civil aircraft causing a human fatality or injury and number of injuries and fatalities by bird species, USA, 1990–2015 [28].

Species of wildlife	No. of strikes	No. of humans
Strikes causing fatalities		
Unknown bird	6	8
Red-tailed hawk	1	8
American white pelican	1	5
Canada goose	1	2
White-tailed deer	1	1
Brown pelican	1	1
Turkey vulture	1	1
Total fatalities	**12**	**26**
Strikes causing injuries		
Canada goose	15	117
Unknown	46	61
White-tailed deer	19	27
Ducks	17	20
Turkey vulture	15	18
New World vulture	10	10
Ring-billed Gull	3	9
Red-tailed hawk	7	9
Gulls	8	9
Bald eagle	4	7
Geese	7	7
Mallard	5	6
American kestrel	1	5
Several birds including Hawks, golden eagle, American coot, cattle, Osprey, etc.	72	95
Total injuries	**229**	**400**

The registered number of bird strikes was 198 in 1988, 120 in 1989, and 81 in 1990. Strikes causing no damage to aircraft are not counted in these statistics.

Table 4.7 presents the distribution of bird strikes by height [30].

From Table 4.7 it is possible to conclude the following:

- The biggest number of strikes are in the aerodrome (0–100 m). Its annual percentage varied from 52% in 1988 to 81% in 1990.
- For heights between 0 m and 400 m, the annual bird strike varied from 75% in 1988 to 94% in 1990.

Table 4.8 lists the percentages of bird strike for different phases of flight for a three-year period (1988–1990). In all three years, bird strikes have the greatest percentage

Table 4.5 Minimum projected annual losses in aircraft downtime (hours) and repair and other costs (inflation-adjusted US$) from wildlife strikes with civil aircraft, USA, 1990–2015. Losses are projected from mean reported losses per incident [28].

Year	No. of adverse incidents	Down-time (hr)	Repair costs (US$ 1 million)	Other costs (US$ 1 million)	Total costs (US$ 1 million)
			Minimum projected losses		
1990	424	23 892	92	26	118
1995	655	63 052	339	148	487
2000	1112	217 046	111	129	241
2005	975	85 550	263	76	339
2010	1128	74 777	145	15	160
2015	1451	69 497	203	27	229
Total (1990–2015)	**24 478**	**2 925 926**	**4062**	**909**	**4971**
Mean (1990–2015)	**941**	**112 536**	**156**	**34**	**191**

Table 4.6 Number of civil aircraft with reported damage resulting from wildlife strikes, USA, 1990–2015 [28].

Damage category	26 years	Total (%)
	Minimum projected losses	
None	104 314	63
Unknown	46 572	28
Damage	13 558	8
Minor	7230	4
Uncertain	2831	2
Substantial	3460	2
Destroyed	37	<1
Total	**164 444**	**100**

during approach. The climb phase has the second worst percentage. Landing roll is followed, while takeoff run experienced the lowest percentages.

When looking at which bird species are responsible for accidents, it is recorded that gulls and terns are responsible for 35% of strikes, followed by predominantly perching birds causing 17%. Pigeons and doves are responsible for 16% of bird strikes. The two groups of hawks/eagles and ducks/geese are each responsible for 10% of these accidents. Finally, the three groups of crows, rooks, owls, and lapwings, and plovers are responsible for a total of 11% [30].

Table 4.7 The distribution of bird strikes by height in Russia (1988–1990) [30].

Height	Year			
	1988	1989	1990	Total
0–100 m	52 (58%)	25 (52%)	30 (81%)	107 (61%)
101–400 m	15 (17%)	12 (25%)	5 (13%)	32 (18%)
401–1000 m	11 (12%)	11 (23%)	1 (3%)	23 (13%)
More than 1000 m	12 (13%)	—	1 (3%)	13 (8%)
Total	90 (100%)	48 (100%)	37 (100%)	175 (100%)
Unknown	108	72	44	224

Table 4.8 Distribution of bird strike by the phase of flight in Russia (1988–1990) [30].

Phase of flight	Year			
	1988	1989	1990	Total
Approach	32 (41%)	23 (42%)	12 (27%)	67 (37%)
Climb	20 (26%)	16 (29%)	8 (18%)	44 (25%)
Landing roll	13 (16%)	5 (9%)	9 (20%)	27 (15%)
Takeoff run	3 (4%)	7 (13%)	11 (24%)	21 (12%)
Other	10 (13%)	4 (7%)	5 (11%)	19 (11%)
Total	78 (100%)	55 (100%)	45 (100%)	178 (100%)
Unknown	120	65	36	221

Statistics for the parts of aircraft frequently hit by birds is recorded in Table 4.9. In the three years (1988–1990), the statistics outlines that:

- the engine is the most critical part, where bird strikes varied from 39% in 1988 to 49% in 1990;
- the landing gear was the least struck part, with the number of strikes decreasing from 4% in 1988 to 1% in 1990.

4.7 Military Aircraft

4.7.1 Introduction

Most military aircraft use the same low altitude airspace as large concentrations of birds. Consequently, reporting of bird strikes and planning corrective actions becomes an urgent necessity.

In recent years, a large but unknown number of military aircraft have crashed as a result of bird strikes and many aircrews have been killed [29]. In the USA alone, the

Table 4.9 Part damaged of aircraft due to a bird strike in Russia (1988–1990) [30].

Part struck	Year			
	1988	1989	1990	Total
Engines	74 (39%)	52 (41%)	42 (49%)	169 (42%)
Wings	53 (26%)	34 (27%)	20 (23%)	107 (27%)
Windshield	7 (4%)	5 (4%)	5 (6%)	17 (4%)
Fuselage	4 (2%)	10 (8%)	1 (1%)	15 (4%)
Radome	17 (9%)	9 (7%)	7 (8%)	33 (18%)
Headlights	11 (16%)	8 (6%)	5 (6%)	24 (6%)
Landing gear	8 (4%)	1 (1%)	1 (1%)	10 (2%)
Other	14 (8%)	8 (6%)	5 (6%)	27 (7%)
Total	188 (100%)	127 (100%)	86 (100%)	401 (100%)

USAF and Navy/Marine Corps experience at least 3000 bird strikes, causing more than $75 million in damage, every year. However, only 20% of actual bird strikes are reported.

Highlights of bird strike accidents are described in the following sections, compiled from many sources [8, 31–36]. Military statistics are not easily available in open sources. Many accidents are considered classified for some 20–30 years. Moreover, reporting of military aircraft accidents was not strictly requested for the early decades of flight. For both reasons, less data are available compared with civilian accidents and much of the available data is rather old and belongs to some 20–30 years ago.

4.7.2 Statistics for Military Aircraft Accidents

Going back to the Second World War, there were at least six fatal accidents in the South-East Asia area for the aircraft of UK RAF due to bird strikes, with the planes including Hurricane, Spitfire, and Mosquito [7]. Also, the USAF in the Pacific suffered more damage due to bird strikes than to enemy action. The first fatal accident to an RAF jet-powered aircraft is thought to be a Canberra twin-engine bomber in 1953. Many more RAF aircraft were lost between 1950 and 1979 (and earlier) due to bird strikes.

In the USA, during the period 1968 to 1984, 16 aircraft were destroyed, and 13 military pilots were killed because of bird strikes [35]. The number of recorded bird strikes with military aircraft are increasing, due to better records being kept and because aircraft are spending more time at the lower altitudes where strikes occur. As aircraft become more sophisticated and valuable, it is imperative that bird strikes be minimized to prevent aircraft damage or pilot injury. Costs for 1982 came to $14 million, which included the loss of an F16 aircraft. Moreover, in the early 1980s, west European military forces may have been losing up to 10 jet fighters per year due to bird strikes [36].

From 1950 to 2000, there were some 131 accidents due to bird strike in the following countries: Austria, Canada, Denmark, Germany, Netherlands, Norway, Sweden,

Switzerland, UK, and the USA. These accidents are listed in Table 4.10 [36], and were distributed as follows:

- 69 in Europe;
- 9 in Canada;
- 32 in the USA;
- 5 elsewhere; and
- 16 at unknown locations.

Most involved fighters, attack aircraft, and training aircraft, but two accidents involved four-engine bombers and patrol aircraft. The largest number of accidents were during high-speed, low-level flight (altitude less than 1000 ft AGL). Most of these accidents involved engine ingestions and windscreen penetrations. In Europe, gulls were first while buzzards were second in causing bird strike events. In the USA, vultures were the most serious problem.

From Table 4.10, it can be seen that loss rates (aircraft per year) have declined over the five decades (1950–2000) in countries such as Germany, Sweden, and Canada. This is due to the retirement of Canadian and German F104 aircraft, which had high loss rates, as well as reductions of total fleet size and flying hours in other countries. In the USA, however, the loss rates have not declined since F-16 and T-38 aircraft, in particular, continue to be lost in bird strikes at the rate of one to two per year [36].

Concerning fatalities, there were at least eight aircrew members killed in the 1960s, 17 in the 1970s (excluding RAF), 18 in the 1980s, and two in the 1990s [36]. The worst single accident was the USAF B-1B lost due to a pelican strike in 1987, with three fatalities.

The crew manage to eject in a high proportion of the serious accidents, and most ejections are successful.

Table 4.10 Minimum numbers of military aircraft of 10 counties lost due to birds (the 1950s–1990s) [36].

Country	1950s	1960s	1970s	1980s	1990s	Total
Europe						
Denmark	0	0	0	0	0	0
Austria	0	0	0	0	0	0
Germany	?	5	11	6	0	22
Netherland	2	3	2	2	1	10
Norway	?	?	1	2	?	3+
Sweden	?	2+	7	0	?	9+
Switzerland	0	0	1	0	0	1
United Kingdom	+	+	+	13	2	15+
North America						
USA	0	11	14	15	11	51
Canada	0	10	4	2	1	17
Total	2	32	42	40	15	131

Some bird strikes related accidents have occurred outside the borders of the operating countries. For example, almost half of the known Canadian losses to bird strikes were in Europe, and the USAF lost three aircraft to bird strikes on the Bardenas Range in Spain during the 1980s.

Serious bird strike-related accidents have occurred at all times of the year in both Europe and North America. However, in Europe, serious accidents seem most common in spring (March to April) and late summer /early autumn (July to October). The limited data from Canada may reflect a similar pattern, but with the spring peak delayed to May. USAF statistics on the seasonal occurrence of bird strikes show peaks in spring and late summer/autumn [37].

In another study for the same period (1950–1999) but covering 32 countries [33], rather than the 10 countries discussed above, a total of 286 serious bird-related accidents to military aircraft were listed instead of the 131 identified above. The countries considered include most of those in Europe (east to Russia), Canada, USA, Israel, Australia, and New Zealand.

Western and Northern Europe had a total of 60 fatalities in the period 1950–1999. The greatest fatalities were listed in Belgium (37 in 55 years and 34 in the 1990s alone), Netherlands had two fatalities, while both of Norway and Portugal had one fatality each. Each of Sweden and Spain had five fatalities while the UK had nine fatalities [33].

Eastern Europe recorded a total of 22 fatalities. The Czech and Slovak Republics (former Czechoslovakia), plus Hungary, and Albania each recorded one fatality. East Germany encountered six fatalities while the Former Soviet Union had 13 fatalities [33].

North America recorded 53 fatalities. Canada had two fatalities while the USA had 51. Of the other counties, Israel and Australia each recorded three fatalities, giving a total of six.

Of the 286 serious bird-related accidents considered above, eight crashes – six to British aircraft and two to US aircraft – were due to low-altitude maneuvers to avoid collisions with birds [33].

A detailed study on UK military accidents over 82 years (1923–2004) reports that among the 108 losses, 101 were due to bird strikes while seven crashes were encountered during attempts to avoid birds [34]. These 108 accidents are still underestimates of the number of actual strikes. This is attributed to two reasons: records for early years are incomplete and losses for which bird strike evidence is lacking or weak are omitted. Of these 108 serious bird-related accidents, 63 were in or near the UK, 12 in were in continental Europe, 23 in South and South-East Asia, 4 in the Mideast/SW Asia region, 4 in Africa, 1 in the Falklands, and 1 was unknown. At least 25 aircrew members were killed.

Figure 4.40 illustrates both aircraft lost and fatalities in the UK during the period 1923–2004.

In the 1940s, 33 accidents occurred. Most of those (28) occurred during World War II, despite the difficulty in determining the causes of many accidents during wartime. The highest annual numbers of known losses due to bird strike were 7 or 8 aircraft per year in each of the three years 1943–1945. The known-loss rate decreased rapidly after mid-1945 with demobilization. The average known-loss rates for 1950–1979 were 1.3 aircraft per year which dropped to 0.8 in 1980–2004.

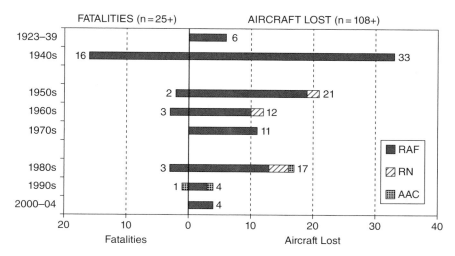

Figure 4.40 Number of known serious bird-related accidents (right) and associated human fatalities (left) in UK military aviation. *Source:* Courtesy IBSC [34]. RAF is Royal Air Force, RN is Royal Navy, and AAC is Army Air Corps.

4.7.3 Statistics for Ex-Soviet Union Air Force in East Germany

Detailed statistics for the ex-Soviet Union Air Force in East Germany over two periods (1970–1981 and 1985–1991) is summarized here [38].

Concerning the different heights above ground and the percentages of bird strike accidents, a summary is given:

- 0–100 m (39.5%)
- 101–300 m (26.6%)
- 301–500 m (14.9%)
- 501–1000 m (12.6%)
- 1001–2000 m (5.5%)
- >2000 m (0.9%)

Bird strike accidents for the ex-Soviet Union Air Force in East Germany [38] varied between different categories of military aircraft for the same flight zone. Table 4.11 presents how these percentages differ for fighters and bombers. The flight phases

Table 4.11 Distribution of bird strike by the phase of flight in ex-Soviet Union Air Force in East Germany during 1970–1981 and 1985–1991 [38].

	Phase of flight							
Aircraft type	Taxi	Takeoff	On route	Zone	Circle	Firing ground	Landing	Landing/takeoff
Fighter	18	33	46	14	42	7	108	3.2
Bomber	–	1	9	1	–	1	3	3.0

Table 4.12 Part damaged of aircraft due to bird strike in Ex-Soviet Union Air Force in East Germany during (1970–1981) and (1985–1991) [38].

				Part struck			
	Engine	Wing	Nose fuselage (Radome)	Landing gear	Windshield	Tail	Unknown nil damage
Number of strikes	263	39	17	33	24	3	76
Numbers of strikes without damage	83	16	12	28	15	–	76
%	58	8.5	3.6	7.2	5.3	0.7	16.7

considered are taxi (rolling start and finish), takeoff, landing, on the circle, in the zone, on the route, firing ground. The following explanations are provided.

- Circle flight phase consists of takeoff, gaining height to 500–600 m, circle flight and landing at the same place where takeoff took place (duration is nearly 10 minutes).
- Flight in the zone consists of different pilotage figures over a pre-determined route and altitude (duration 3–35 minutes).
- Bombardments and rocket shooting are carried out at the firing ground from different heights with or without diving on ground targets.

The part of the aircraft damaged due to bird strike [38] is given in Table 4.12. In 76 cases, the place struck has not been mentioned, except for "bird strike" or "nil damage."

Variations of bird strike accidents according to time of day and weather conditions for fighters and bombers are given in Table 4.13 [38]. The following abbreviations are use:

- day (D);
- night (N);
- simple weather conditions (S) determined as good visibility, high clouds, or clear sky;
- complicated weather conditions (C) are related to low unbroken cloud and a low visibility.

Based on the time of year, bird strike accidents showed two peaks [38]. The first peak of collisions took place during spring migration in March/April, while the second peak was in July/August.

Table 4.13 Bird strike distribution based on time and weather conditions [38].

Aircraft type	DS	DC	NS	NC	$\left(\dfrac{N}{D+N}\right)\%$
Fighter	205	90	56	46	25.7
Bomber	13	3	6	1	30.4

Finally, the most dangerous birds, which collided with ex-Soviet Union Air Force planes in East Germany [38] were duck, goose, mute swan, birds of prey, partridge, pheasant, gull, lapwing, owl, heron, pigeon, swift, thrush, crow, starling, lark, swallow, and sparrow.

4.7.4 Details of Some Accidents for Military Aircraft

28 September 1987

A B-1B bomber (Figure 4.41) crashed into farmland in southern Colorado after striking a flock of birds. Birds were sucked into its engines. It lost two of its four engines; one in a fire. Three crewmen were killed, but three others parachuted to safety. It was the first crash of a production model of the US$ 270 million B-1B. The three survivors were picked up by a La Junta police officer not far from the crash site and were reported in good condition with minor injuries at the USAF Academy Hospital in Colorado.

22 September 1995

A Boeing E-3B Sentry 77-0354 with callsign Yukla 27 (a military Boeing 707 derivative equipped with airborne warning and control system (AWACS) system) operated by the USAF was on a training mission [39]. At 07:45 the aircraft was cleared for takeoff and the throttles were advanced. As the plane rotated for lift-off, numerous Canada geese were ingested in the #1 and #2 engines, resulting in a catastrophic failure of the #2 engine and a stalling of the #1 engine. The pilots still tried to take off, initiated a slow climbing turn to the left, and began to dump fuel. The plane could not get enough power and plunged into a heavily wooded area of Fort Richardson, less than a mile from the runway, broke up, exploded, and burned. All the four crew and 20 passengers were killed (Figure 4.42).

Figure 4.41 B-1B bomber. *Source:* https://en.wikipedia.org/wiki/Rockwell_B-1_Lancer#/media/File:B-1B_over_the_pacific_ocean.jpg.

Figure 4.42 1995 bird strike accident for E-3 Sentry AWACS [66]. *Source:* https://en.wikipedia.org/wiki/1995_Alaska_Boeing_E3_Sentry_accident#/media/File:AWACS_Engine_(ADN).jpg.

Figure 4.43 NATO AWACS E-3A accident. *Source:* Courtesy US Navy [58].

14 July 1996

The pilot of a NATO AWACS E-3A (Boeing E-3A Sentry 707-300B) aborted the takeoff from Préveza-Aktion Airport (PVK), Greece, when he assumed from the sounds he heard that they had bird ingestion [40]. The aircraft could not be brought to a stop on the runway, and thus overran the runway and dipped into the sea. The fuselage broke into two, destroying the aircraft, but there were no casualties among the 16 crew members on board (Figure 4.43). However, no evidence of a bird strike was found.

15 July 1996

A Lockheed C-130H Hercules CH-06 (Figure 4.44) of the Belgian Air Force was descending to Eindhoven Airport when, possibly because of the presence of a large

Figure 4.44 Crash of Lockheed C-130H Hercules CH-06 of the Belgian Air Force due to bird strike [67].

number of starlings near the runway, the co-pilot initiated a go around. Both left-hand engines (#1 and #2) lost power due to bird ingestion. However, for unknown reasons, the crew shut down engine #3 and feathered the propellers. Thus, the plane turned left and crashed off the left side of the runway. Within seconds a fire erupted which was fed by the oxygen from the airplane's oxygen system. All the four crew and 34 passengers were killed [41].

10 July 2006

A Russian Tupolev TU-134 military passenger plane, operated on behalf of the Black Sea Fleet Aviation unit (VVS ChF) of the Russian Navy, sucked in a flock of birds into the left engine on takeoff from Sevastopol, Ukraine. The takeoff was aborted, but the aircraft overran the runway and burst into flames [42]. The aircraft crashed near Simferopol. Three navy officers were injured while all 20 passengers survived (Figure 4.45).

18 January 2007

A twin engine training jet, USAF T 38 (Figure 4.46) was in a two-ship low-level navigation training exercise. It encountered birds (mallard ducks), which were ingested into the engine and struck nose. The aircraft crashed about 40 miles south of Memphis, TN, 10 miles WNW of Batesville, MS, in a swampy area of western Panola County. Both the pilot and the trainee pilot ejected safely.

(a)

(b)

Figure 4.45 Crash of TU-134 of the Russian Navy military passenger plane. *Source:* Courtesy Flight International.

21 April 2008

An American F-111 bomber was at about 900 m altitude on a test bombing raid at Evan's Head, NSW, Australia, when a pelican bounced off its fiberglass nose before being sucked into an engine. The two RAAF crew managed a skillful recovery and landing. The bird strike caused damage to the nose and radome (Figure 4.47), a hole in one wing and damaged one engine. Neither of the two RAAF crew was injured [43].

26 June 2013

A USAF F16D Fighting Falcon struck a bird while conducting a military training "touch and go." The aircraft crashed, but both pilots ejected near Luke Air Force Base in Arizona [44].

Figure 4.46 Twin engine training jet, USAF T 38 [68]. *Source:* https://en.wikipedia.org/wiki/Northrop_T-38_Talon#/media/File:T-38_Talon_over_Edwards_AFB.jpg.

Figure 4.47 American F-111 bomber struck by pelican bird causing nose damage [69].
Source: Courtesy US DoD, Avian Hazard Advisory System (AHAS). http://www.usahas.com/AHAS2015/gallery/image.aspx?image=F111%20Bird%20Strike.

3 May 2018

A Russian fighter aircraft Su-30SM collided with a bird (undefined) during its initial climb at 09:45 local time. The aircraft crashed into the sea off the city of Jiblah, Latakia province, Syria. Both crew members were killed [45].

Finally, Table 4.14 summarizes the number of bird strikes with military aircraft in three years in the USA.

4.7.5 Details of Accidents for Military Aircraft in Norway in 2016

Most countries now have their own system for reporting bird strikes for both civil and military aircraft. These data are saved in files and exchanged between domestic

Table 4.14 Bird strikes by aircraft in the USA (1980–1982) [35].

Aircraft	No. of bird strike	No. of flight hours	Bird strike rate (per 1000, 000 hr)
B-52	353	385 928	91.5
C-130	266	1 095 888	24.3
E-3	15	66 884	22.4
T-43	17	52 530	32.4
F-15	132	390 547	33.8
F-16	117	188 979	61.9
T-37	265	897 344	29.5
T-38	394	1 039 945	37.9

authorities in different countries. As an example for a European country, Table 4.15 lists all details for accidents encountered by aircraft of the Royal Norwegian Air Force aircraft during 2016 [46].

Table 4.15 Bird strikes 2016 in Royal Norwegian Air Force [46][a].

No.	Date (dd/mm/year)	Month	Location	Aircraft	Squadron	Altitude (ft AGL)
1	1/18/2016	1	7119N 2440E; 20NM north of Hammerfest (Hjelmsøybanken)	P-3	333	1700
2	4/12/2016	4	ENOL	F-16	338	low
3	4/23/2016	4	6122N 0201E; North Sea	Sea King	330	100
4	6/2/2016	6	ENGM	DA-20	717	750
5	6/22/2016	6	Bamako, MALI	C-130	335	10
6	6/23/2016	6	ENOL	F-16	338	50
7	7/15/2016	7	Bamako, MALI	C-130	335	2000
8	8/1/2016	8	ENOL	F-16	338	0
9	8/24/2016	8	ENBO	F-16	331	low
10	8/29/2016	8	ENOL	F-16	338	low
11	8/30/2016	8	ENBO	F-16	331	0
12	9/5/2016	9	Bamako, MALI	C-130	335	400
13	9/21/2016	9	ENBO	F-16	331	low
14	10/11/2016	10	32V NL 295800, Østfold county	Bell 412	339	100
15	10/21/2016	10	Bamako, MALI	C-130	335	<1000
16	10/25/2016	10	15 NM NW of Osensjøen, Hedmark county	C-130	335	3900 msl
17	10/26/2016	10	5,9 NM final Bamako, MALI	C-130	335	3300 msl

Table 4.15 (Continued)

No.	Speed (knot)	Phase of flight	Bird species	# Bird hit	Bird size	Part1 struck	P1 damaged	Part2 struck
1	220	ll, er	2 seagulls	2	large	engine #1	no	tail
2	u	to	Brambling	1	small	cannon system	no	
3	100	ll, er	seagull	1	medium	rotor	no	
4	u	er	unknown	1	unknown	leading edge	no	
5	115	la	unknown medium size	1	medium	radome	no	nose
6	230	to	shorebird or passerine bird	1	small	air intake	yes	engine
7	200	app	bat	1	medium	nose	ja	air intake
8	170	t&g	Crow	1	medium	wing	nei	
9	u	la	Ruff	1	medium	underwing stores	unknown	
10	180	to	8 Golden plovers	8	medium	vintage	unknown	air intake, engine, fuselage
11	100	to	unknown small	1	small	wing	unknown	
12	u	app	unknown small	1	small	nose	unknown	
13	u	la	Meadow pipit	1	small	underwing stores	no	
14	80	ll, er	unknown	1	unknown	nose	unknown	
15	120	app	unknown medium size	1	medium	propeller	no	
16	210	ll, er	unknown small	1	small	nose	no	
17	220	app	swallow/ martin	1	small	air intake	no	

No.	P2 damaged	Damage	Comments	Effect on flight
1	no	NO	just a small dent in the engine	returned to base
2		NO		none
3		NO		landed nearest base: ENFB
4		NO	just a small dent in leading edge	unknown

(Continued)

Table 4.15 (Continued)

No.	P2 damaged	Damage	Comments	Effect on flight
5	no	NO		none
6	no	YES	the oil cooler was replaced	none
7	no	YES	damage to fuselage	landed nearest base: GABS
8		NO	feather on leading edge	returned to base
9		u	bird remains	none
10	unknown	YES		returned to base
11		u		canceled departure
12		u		none
13		NO		none
14		u		returned to base
15		NO	jackdaw, small crow	landed nearest base: GABS
16		NO		returned to base
17		NO		none

a) Table 4.15 is extracted reference [46] through private communication with Dr. Christian Kierulf Aas, Natural History Museum, Oslo, Norway.

Abbreviations: ll, er = low level, en route; to = take off; er = en route; la = landing; app = approach; t&g = touch and go; u = unknown.

4.7.6 Comparison between Bird Strikes with Civilian and Military Aircraft

As a conclusion, Table 4.16 presents a comparison between bird strikes with civilian and military aircraft.

4.8 Helicopters

4.8.1 Introduction

A review is given here for the available wildlife strike records about rotary-wing aircraft (helicopters) from the following four US military services [52]:

- US Army [ARMY] during 1990–2011;
- USAF during 1994–2011;
- US Navy and US Marine Corps [NAVY] during 2000–2011;
- US Coast Guard [USCG] during (1979–2011).

Comparison of bird strikes between fixed-wing aircraft and helicopters reached the following conclusions [8, 53–55]:

(1) helicopters were more likely to be damaged than fixed-wing aircraft;
(2) windshields on helicopters were more frequently struck and damaged than windshields on fixed-wing aircraft;
(3) bird strikes with helicopters led to injuries to crew or passengers more than with fixed-wing aircraft.

Table 4.16 Comparison of bird strikes with civilian and military aircraft.

	Civil aircraft	Military aircraft
Incidents	169 856 (1990–2015) [28] Number of bird strikes for US aircraft in 2015 are 13 795 and damaging strikes 616	95 383 (1985–2011), [47] USAF incur 2500 bird strikes annually [48]
Fatalities	276 (1912–2012) 26 (1990–2015)	141 fatalities (1950–1999) 33 fatalities (1985–2011) one human death per 2000 strikes [49]
Serious accidents	>55 (1912–2012)	286 (1950–1999)
Aircraft completely damaged	108 (1912–2012)	39 (1985–2011)
Impacted modules	engines and windshield	windshields and engine
% of reporting	Some 50%	20–30% of accidents
Vulnerability to strikes	Less	More
%Strikes during the takeoff phase	90%	54%
Critical altitude	< 1500 m	<300 m
Birds causing general aviation accidents	Transport aircraft and executive jets [7]: gulls/lapwings, pigeons, birds of prey, perching birds, waterfowl Aircraft of 5700 kg and less: birds of prey, geese, duck, gulls, pelican, cormorant, other water birds, jackdaw	Altitudes <500 ft [35]: gulls, terns, pigeons, doves, waterfowl, and birds of prey Altitudes > 500 ft waterfowl, gulls, terns, vultures, passerines, birds of prey
Costs	Annually US$ 1200 M worldwide US$ 800 M	Total: US$ 820 M (USAF up to 2012) [50] Annually USAF US$ 35 million (1985–1998) [51]

APHIS-Wildlife Services (WS) researches that included data from all four military branches and the Federal Aviation Administration's National Wildlife Strike Database reached the following findings and suggestions [53–55]:

- For nearly 4500 bird strike incidents to helicopters, the majority were during autumn (Sept–Nov; 41.6% of all strikes), and its least was during winter (Dec–Feb; 10.4%).
- Bird strikes with civil and military helicopters resulted in 61 human injuries and 11 lost lives since 1990–2013.
- The average cost of a damaging strike of military helicopters ranged from US$ 12 184 to US$ 337 281 per incident.
- Although bird strikes to military helicopters occurred during all phases of aircraft flight, strikes occurred most frequently when aircraft was traveling en route or engaged in terrain flight.
- The windscreen is the most frequent impact location for reported bird strikes for all civil and military helicopters.

Figure 4.48 A typical bird strike with a helicopter. *Source:* Courtesy USDA-APHIS [52].

- Bird strikes to military helicopters resulted in a total of 8 human injuries (typically cuts, lacerations, and bruises to pilots and co-pilots).
- Most damaging bird strikes with helicopters are due to raptors and vultures.
- The statistics for bird strikes with military helicopters operating overseas in more than 31 foreign countries outlined that two-thirds of overseas bird strikes to ARMY helicopters occurred during deployments in the Middle East, whereas overseas bird strikes to USAF helicopters occurred most frequently in Afghanistan and the Middle East. Larks, perching birds, doves, and pigeons were struck most often. Bird strikes to military helicopters are both costly and deadly.
- There has been a continuous increase in bird strike reporting, and an increase in the number of flight plans modified to avoid or minimize low-level flights over or near landfills and other known wildlife attractants.
- The sections of helicopters most commonly struck and damaged (such as windscreens and main rotor system) may need to be reinforced for a better withstanding the impact of strikes with large birds.

Figure 4.48 illustrates a typical bird strike with a helicopter.

4.8.2 Statistics for Bird Strikes with Civil and Military Helicopters in the USA

Number of strikes for civil and military helicopters in the USA
Table 4.17 illustrates the number of all reported and damaging wildlife strikes for civil and military helicopter aircraft in the USA [52]. Data for civil helicopters is listed for the period 1990–2011, while data for military aircraft are listed for different helicopter groups both in the USA and overseas flight operations in the period 1979–2011 [52]. Table 4.18 represents the annual number of reported bird strikes with civil helicopters in the FAA's National Wildlife Strike Database (NWSD) [52].

Table 4.17 Number of all reported and damaging wildlife strikes for civil and military helicopter aircraft in the USA [52].

Category (Location)	No. of reported strikes	No. of reported damaging strikes	Damaging strikes (%)
Civil (in USA), 1990–2011	1044	367	35.2
Army (in USA), 1979–2011	318	134	42.1
USAF (in USA), 1979–2011	1071	41	3.9
Navy (in USA), 1979–2011	845	103	12.2
USCG (in USA), 1979–2011	251	102	40.6
Army (overseas), 1979–2011	238	175	73.5
USAF (overseas), 1979–2011	463	30	6.5

Table 4.18 The annual number of reported bird strikes with civil helicopters [52].

Year	1990	1991	1993	1994	1996	1998	1999	2000	2001	2002	2003	2004
No. of reported bird strikes	14	18	13	15	26	27	29	19	33	31	21	39
Year	2005	2006	2007	2008	2009	2010	2011	2012	2013	2014	2015	
No. of reported bird strikes	23	33	46	44	127	162	209	220	240	216	231	

4.8.3 Statistics for Bird Strikes with a Flight Phase

Table 4.19 lists the percentage of reported bird strikes in the USA by the phase of flight for military rotary-wing aircraft for each military service during 1979–2011 [52].

Table 4.19 Percentage of reported bird strikes, by the phase of flight, in the United States for military rotary-wing aircraft during 1979–2011 [52].

Phase of flight	US Army	USAF	Navy	USCG
En route	52.3	31.3	28.6	73.6
Terrain flight	22.2	41.4	25.0	0.6
Hovering	1.8	4.2	4.2	1.9
Approach	14.7	7.2	12.7	8.8
Pattern	–	5.3	7.1	3.8
Landing	1.1	3.3	4.0	3.1
Taxiing	1.1	2.3	5.1	–
Touch and go	–	0.2	2.2	1.9
Takeoff	3.2	2.7	5.5	1.9
Climb out	3.6	2.1	5.6	4.4

Table 4.20 The percentages of bird strike by the time of day [52].

Time of day	US Army	USAF	Navy/USCG
Day	71.7%	27%	50%
Night	28.9%	73%	50%

4.8.4   Statistics for Bird Strikes with Time of Day

Table 4.20 identifies the percentages of bird strikes by the time of day [52].

4.8.5   Statistics for Parts of Helicopters Struck by Birds (January 2009 Through February 2016)

Bird strike events on Part 27 rotorcraft were reported on the following components. The total number of strikes is 1614 [52]:

- 584 on the windshield, of which 142 had damaged the windshield;
- 107 were denoted as penetration;
- 366 on the main rotor;
- 235 on the nose/radome;
- 181 on the fuselage;
- 53 on the tail rotor or empennage;
- 22 on the landing gear;
- 21 on the engine(s), with 7 reported as ingested by the engine;
- 4 on lights with no other component impacted;
- 148 on other components.

4.8.6   Statistics for Bird Species Striking and Damaging Helicopters

Table 4.21 identifies the number of most critical bird strikes and damaging strikes for different categories of military rotary-wing aircraft (helicopters) during 1990–2011 [52]. Other wildlife groups are listed by Washburn et al. [52].

4.8.7   Fatal Accidents

A brief description of some accidents encountered by helicopters will be given here [8].

16 April 1965

While a Sikorsky S62A helicopter powered by turboshaft engine GE CT581001 was departing from San Francisco, a homing rock pigeon was sucked into the intake causing a compressor stall. This resulted in a retreating rotor blade stall which next collided with wires/poles. Fortunately, neither of the two occupants were killed or injured.

29 January 1983

A Bell 47 helicopter was flying at a speed of 45 knots and an altitude of 15 ft above the water when a bird came through the door opening and hit the pilot's right temple. He lost control and the helicopter crashed into the sea. The passenger suffered minor injuries [8].

Table 4.21 The number of most critical bird strikes and damaging strikes for different categories of military helicopters during 1990–2011 [55].

Birds group	Army All strikes	Army Damaging strikes	USAF All strikes	USAF Damaging strikes	Navy All strikes	Navy Damaging strikes
Blackbirds and starlings	—	—	7	—	1	—
Doves and pigeons	1	—	19	3	8	—
Larks	—	—	19	—	—	—
Gulls	1	1	2	—	21	2
Perching birds	—	—	17	1	6	2
Raptors and vultures	5	2	3	2	16	4
Shorebirds	—	—	6	1	12	1
Sparrows	—	—	17	—	2	—
Thrashers and thrushes	—	—	20	—	1	—
Warblers	—	—	26	—	1	—
Waterfowl	3	2	4	—	3	—

20 January 1985

While a Hughes 369 helicopter was flying at an altitude of nearly 400 ft and nearly 0.5 mile offshore, the pilot was unable to avoid a large flock of birds. An extreme vibration developed, so he ditched the helicopter, and it sank. The pilot swam and safely reached the shore [8].

27 January 2000

A Bell 407 Long Ranger was about 25 nautical miles SW from Panama City, Panama, at an altitude of 1500 ft and speed of 90 knots, when a black vulture (1.7 kg) penetrated the windshield and struck the pilot, knocking him unconscious. The co-pilot attempted to take over, but the helicopter crashed onto a hillside and rolled down a slope. The pilot and one passenger were killed, and the co-pilot and two passengers were seriously injured [8].

19 March 2000

A Bell 212 helicopter was cruising at about 400–500 ft when it encountered a large bird and crashed. The wreckage trail was nearly 1000 ft long and included a 12 ft section of the main rotor blade and the complete right windshield, but the left windshield was not found. The accident resulted in one fatality [8].

28 October 2003

An Heli Inter Guyana AS 350 Ecureuil helicopter was flying over jungle when it encountered a bird strike. The bird entered the cabin through the left-hand windscreen, leading to the left rear door opening. A passenger fell from the helicopter and was killed [8].

Figure 4.49 Wreckage of PHI Sikorsky S-76C. *Source:* Courtesy NTSB [8].

Figure 4.50 US Air Force HH-60G Pave Hawk helicopter. *Source:* Courtesy NTSB [8].

4 January 2009

A Sikorsky S76C helicopter operated by Petroleum Helicopters was ferrying oilfield workers to an oil platform in the Gulf of Mexico. A bird struck the canopy just above the top edge of the windshield. This resulted in of both the left-hand and right-hand sections of the windshield shattering. Feather and other bird remains collected from the canopy and windshield identified that the bird was a female red-tailed hawk (1.1 kg). The accident resulted in eight fatalities and a seriously injured person [8]. Figure 4.49 illustrates the wreckage of the helicopter.

7 January 2014

A USAF HH-60G Pave Hawk helicopter crashed as a result of multiple bird strikes (Figure 4.50). The crash killed four men [56]. The US Accident Investigation Board found that the accident was caused by geese flying through the aircraft's windshield, knocking the pilot and co-pilot unconscious. They were then unable to react when another bird hit the helicopter's nose, disabling stabilization systems and eventually putting the aircraft in an uncontrolled and eventually fatal roll. Only three seconds elapsed between the initial bird strike and the helicopter's crash. No civilians were injured during the crash. The estimated cost of the accident to the US government was US$ 40 million.

23 November 2015

During takeoff of a Eurocopter Airbus AS 350B3 Ecureuil from Jammu, India, a big bird hit the helicopter, which immediately lost control and crashed. A dead vulture was found near the helicopter. The helicopter caught fire as it was descending, apparently as the rotors got tangled in overhead wires.

3 February 2017

During a night training flight, a Swedish Armed Forces Agusta 109LUHs collided with a Western capercaillie (otherwise known as a wood grouse). The bird went through the right-hand windshield and struck the student pilot, resulting in serious facial injuries. The instructor was able to take control and perform a safe landing.

4.9   Conclusions

Aircraft may be destroyed, and occupants may be killed or injured, in accidents due to:

- striking birds
- attempting to avoid birds.

Although not a major cause of accidents, bird strikes may be a serious safety and economic hazard. Remedial measures and tougher aircraft/engines appear to have improved airliner safety. However, replacing aircraft with four-engines with aircraft with two engines imposed a greater risk of bird ingestion in all engines. Engine damage is the major risk for this group of aircraft. Flocking gulls cause some 39% of the accidents and for the major threat. This underlines the importance of thorough aerodrome bird control measures.

Business jets appear to be vulnerable particularly when operated from aerodromes with little or no bird control measures.

Early Russian aircraft operating from "remote" areas where bird control measures are unlikely is the critical group in recent years.

"General aviation" aircraft are most vulnerable to the windshield holing, which is the cause of 51% of the accidents. Birds of prey caused half of the accidents, followed by waterfowl with 34%. General aviation aircraft mostly fly at heights where hard to spot birds are most prevalent.

A high proportion of helicopter accidents were due to the windshield holing, sometimes by heavy birds. Note that helicopters tend to operate at low altitudes where most birds fly. With the present tendency of using faster and quieter helicopters, birds will have less time for avoiding them.

References

1 Brotak, E. (2018). When Birds Strike, Aviation History Magazine. https://www
  .historynet.com/when-birds-strike.htm (acessed 14 January 2019).
2 Thorpe, J. (1996). Bird Strikes to Airliner Turbine Engines, WP63 IBSC 23, London.
3 Kalafatas, M.N. (2010). *Bird Strike – The Clash of Boston Electra*. Brandeis
  University Press.
4 Bird Strike Damage & Windshield Bird Strike, ATKINS Final Report, 5078609-rep-03,
  Version 1.1. https://www.easa.europa.eu/system/files/dfu/Final%20report%20Bird
  %20Strike%20Study.pdf (accessed 29 December 2018).
5 Thorpe, J. (2003). *Fatalities and Destroyed Civil Aircraft Due to Bird Strikes, 1912–
  2002*, IBSC26/WP-9. Warsaw: International Bird Strike Committee.
6 Thorpe, J. (2010). Update on Fatalities and Destroyed Civil Aircraft due to Bird Strikes,
  with Appendix For 2008 & 2009, IBSC29/WP, 29th Meeting of the International Bird
  Strike Committee, Cairns (Australia).
7 John Thorpe, 100 Years of Fatalities and Destroyed Civil Aircraft Due to Bird Strike,
  IBSC30/WP Stavanger, Norway (2012), International Bird Strike Committee. http://www
  .int-birdstrike.org/Warsaw_Papers/IBSC26%20WPSA1.pdf (accessed 2 January 2019).
8 Thorpe, J., Review of 100 Years of Military and Civilian Bird Strikes. https://www.birds
  .org.il/Data/Education/latus%20beshalom%20im%20tsiporim/Migratio%20seminar
  %201999/143_2000762678.PDF (accessed 7 January 2019).
9 Statistics: Cause of Fatal Accidents by Decade http://planecrashinfo.com/cause.htm
  (accessed 7 January 2019).
10 Airbus (2017). A Statistical Analysis of Commercial Aviation Accidents 1958–2016.
  https://flightsafety.org/wp-content/uploads/2017/07/Airbus-Commercial-Aviation-
  Accidents-1958-2016-14Jun17-1.pdf (accessed 7 January 2019).
11 Maragakis, I. (2009). European Aviation Safety Agency, Safety Analysis and Research
  Department Executive Directorate, Bird Population Trends and Their Impact on
  Aviation Safety 1999–2008.
12 Boeing Commercial Airplanes 2017. Statistical Summary of Commercial Jet Airplane
  Accidents: Worldwide Operations | 1959–2017. http://www.boeing.com/resources/
  boeingdotcom/company/about_bca/pdf/statsum.pdf (accessed 14 January 2019).
13 van Es, G.W.H and Smit, H.H. (2019). A Method for Predicting Fatal Bird Strike Rates
  at Airports, NLR-CR-99322. https://www.researchgate.net/publication/265481396_A_
  Method_for_Predicting_Fatal_Bird_Strike_Rates_at_Airports (accessed
  14 January 2019).
14 Aviation Safety Network. https://aviation-safety.net/database/record.php?id=
  19731204-0 (accessed 7 January 2019).
15 Aircraft Accident Report (1976). Report No. NTSB-AAFi-76-19. https://www.fss.aero/
  accident-reports/dvdfiles/US/1975-11-12-3-US.pdf (accessed 7 January 2019).
16 Hawker Siddeley HS 125 Serie 600B G-BCUX: Report on the accident near Dunsfold
  Aerodrome, Surrey, 20 November 1975. https://assets.publishing.service.gov.uk/media/
  5422f0c6ed915d137100030f/1-1977_G-BCUX.pdf (accessed 7 January 2019).
17 Aviation Safety Network, Report into accident on Thursday 20 November 1975. https://
  aviation-safety.net/database/record.php?id=19751120-1 (accessed 14 January 2019).
18 Aviation Safety Network, Report into accident on Tuesday 4 April 1978. https://raviation-
  safety.net/database/record.php?id=19780404-1&lang=nl (accessed 14 January 2019).

19 Aircraft Accident Report -- North Central Airlines, Inc., Convair 580, N4825C, Kalamazoo Municipal Airport, Kalamazoo, Michigan, July 25, 1978. Report Number NTSB-AAR-79-4. http://libraryonline.erau.edu/online-full-text/ntsb/aircraft-accident-reports/AAR79-04.pdf (accessed 7 January 2019).

20 Aviation Safety Network. Learjet 35A HB-VFO Paris-Le Bourget Airport. https://aviation-safety.net/database/record.php?id=19821206-0 (accessed 7 January 2019).

21 Ethiopian Airlines plane crashes. http://www.airsafe.com/events/airlines/eth.htm (accessed 7 January 2019).

22 Cleary, E.C., Dolbeer, R.A., and Wright, S.E. (2006). Wildlife Strikes to Civil Aircraft in the United States 1990–2005. http://digitalcommons.unl.edu/cgi/viewcontent.cgi?article=1006&context=birdstrikeother (accessed 29 December 2018).

23 NTSB National Transportation Safety Board. US Airways Flight 1549, Water Landing Hudson River, January 15, 2009. http://graphics8.nytimes.com/packages/images/nytint/docs/documents-for-the-testimony-of-us-airways-flight-1549/original.pdf (accessed 7 January 2019).

24 Barr, N. (2009). How a flock of birds caused a plane's engine to fail. www.express.co.uk/news/world/131036/How-a-flock-of-birds-caused-a-plane-s-engine-to-fail (accessed 7 January 2019).

25 Hradecky, S. (2011). Accident: Royal Air Maroc B734 at Amsterdam on June 6th 2010, a flock of birds, engine fire. *The Aviation Herald*. http://avherald.com/h?article=42c90189/0000 (accessed 7 January 2019).

26 Preventing Catastrophic Failure of Pratt & Whitney Canada JT15D-5 Engines Following Birdstrike or Foreign Object Ingestion, Safety Recommendation Report, National Transportation Safety Board. https://www.ntsb.gov/investigations/AccidentReports/Reports/ASR1703.pdf (accessed 7 January 2019).

27 Badr Airlines Boeing 737 damaged by bird strike in Sudan, *Wings Hearald*. http://www.wingsherald.com/badr-airlines-boeing-737-damaged-bird-strike-sudan (accessed 7 January 2019).

28 Dolbeer, R.A., Weller, J.R., Anderson, A.L., and Begier, M.J. (2016). Wildlife Strikes to Civil Aircraft in the United States, 1990–2015, FAA report, November 2016. https://wildlife.faa.gov/downloads/Wildlife-Strike-Report-1990-2015.pdf (accessed 2 January 2019).

29 Kitowski, I. (2011). Civil and military Birdstrikes in Europe: an ornithological approach. *Journal of Applied Sciences* 11 (1): 183–191.

30 Ryjov, S.K. (1996). Data of Statistical Study of Bird Strikes with Russian Aircrafts for the period of 1988 to 1990. WP63 BSCE 29,/WP 15, 13–17 May 1996, London.

31 Richardson, W.J. (1994). Serious Bird Strike-Related Accidents to Military Aircraft of Ten Countries: Preliminary Analysis of Circumstances, BSCE 22 / WP21, Vienna, 29 August–2 September 1994.

32 Richardson, W.J. (1996). Serious Bird Strike-Related Accidents to Military Aircraft of Europe and Israel: List and Analysis of Circumstances, BSCE 23 / WP2, London, 13–17 May 1996.

33 Richardson, W.J. and West, T. 2000. Serious Bird Strike-Related Accidents to Military: Updated List and Summary, IBSC 25 / WP SA1, Amsterdam, 17–21 April 2000.

34 Richardson, W.J. and West, T. Serious Bird Strike Accidents TO UK Military Aircraft, 1923 TO 2004: Numbers and Circumstances, IBSC27/WP I-2 Athens, 23–27 May 2005.

35 Payson, R.P. and Vance, J.O. (1984). A Bird Strike Handbook for Base-Level Managers, M.Sc. Thesis, Air Force Institute of Technology, Air University. https://apps.dtic.mil/dtic/tr/fulltext/u2/a147928.pdf (accessed 3 January 2019).

36 Buuma, L.S. (1984). Key Factors determining birdstrike and risks. *International Journal of Aviation Psychology* 2 (t): 91–107.

37 Thompson, M.M, DeFusco, R.P., and Will, T.J. (1986). U.S. Air Force birdstrikes 1983–1985, Proceedings of the Bird Strike Committee, Europe, 18, Pt. l l (Copenhagen): 149–159.

38 V. Jacoby (1998). Analysis of the Bird-Strikes to Ex-Soviet Union Air Force In East Germany, 1970–1981, 1985–1991. IBSC 24/WP 5, Stara Lesna, Slovakia, 14–18 September 1998. http://www.int-birdstrike.org/Slovakia_Papers/IBSC24%20WP05.pdf (accessed 7 January 2019).

39 1995 Alaska Boeing E-3 Sentry accident. https://en.wikipedia.org/wiki/1995_Alaska_Boeing_E-3_Sentry_accident (accessed 7 January 2019).

40 http://www.aviationpics.de/military/1999/awacs/awacs.html (accessed 7 January 2019).

41 Photo of Lockheed C-130H Hercules CH-06. https://aviation-safety.net/photos/displayphoto.php?id=19960715-0&vnr=2&kind=C (accessed 7 January 2019).

42 Tupolev Tu-134. https://aviation-safety.net/database/record.php?id=20060710-1&lang=fr (VIDEO) (accessed 7 January 2019).

43 F-111 aircraft attacked by the collision with Pelicans and the nose wrecked (2008). https://gigazine.net/gsc_news/en/20080421_pelican_strike_f111 (accessed 7 January 2019).

44 Shamim, A. (2013). F-16 Fighting Falcon News: Luke F-16 crashes, both pilots safely eject. http://www.f-16.net/f-16-news-article4748.html (accessed 7 January 2019).

45 Aviation Safety Network, Report into accident on Thursday 3 May 2018. https://aviationsafety.net/wikibase/wiki.php?id=210347 (accessed 14 January 2019).

46 Christian Kierulf Aas (2017). Private communication, Natural History Museum, Oslo, Norway.

47 Sodhi, N.S. (July 2002). Competition in the air: birds versus aircraft, the auk. *A Quarterly Journal of Ornithology* 119 (3): 587–595.

48 Lovell, C.D. and Dolbeer, R.A. (1999). Validation of the United States Air Force Bird Avoidance Model, *Wildlife Society Bulletin*, Vol. 27, Number 1.

49 Neubauer, J.C. (1990). Why birds kill: cross-sectional analysis of U.S. Air Force bird strike data. *Aviation, Space and Environmental Medicine* 61: 343–348.

50 Top 50 USAF Wildlife Strikes by Cost. https://www.safety.af.mil/Portals/71/documents/Aviation/BASH%20Statistics/Top%2050%20USAF%20Wildlife%20Strikes%20by%20Cost.pdf (accessed 14 January 2019).

51 Zakrajsek, E.J. and Bissonette, J.A. (2005). Ranking the risk of wildlife species hazardous to military aircraft. *Wildlife Society Bulletin* 33: 258–264.

52 Washburn, B.E., Cisar, P.J., and DeVault, T.L. (2014). Wildlife Strikes with Military Rotary-Wing Aircraft during Flight Operations within the United States, USDA-APHIS National Wildlife Research Center. https://pdfs.semanticscholar.org/1eb2/edb0475b2b70690539e92b0d02c916aec689.pdf (accessed 14 January 2019).

53 Keirn, G. (2013). Helicopters and Bird Strikes, U.S. Department of Agriculture. https://www.usda.gov/media/blog/2013/06/06/helicopters-and-bird-strikes-results-first-analysis-available-online (accessed 7 January 2019).

54 Bird Strike Hazards and Mitigation Strategies for Military Rotary Wing Aircraft, Project # 11-944. http://www.denix.osd.mil/nr/priorities/birds/bash/fact-sheet-bird-strike-hazards-and-mitigation-strategies-for-military-rotary-wing-aircraft-legacy-11-944 (accessed 7 January 2019).

55 Rotorcraft Bird Strike Working Group: Recommendations to the Aviation Rulemaking Advisory Committee (ARAC) (2017). https://www.faa.gov/regulations_policies/rulemaking/committees/documents/media/ARAC%20RBSWG%20Final%20Report.pdf (accessed 7 January 2019).

56 Learmount, D. (2014). Multiple birdstrike downed Pave Hawk, USAF concludes. https://www.flightglobal.com/news/articles/multiple-birdstrike-downed-pave-hawk-usaf-concludes-401256 (accessed 7 January 2019).

57 National Transportation Safety Board (2010). Loss of Thrust in Both Engines After Encountering a Flock of Birds and Subsequent Ditching on the Hudson River US Airways Flight 1549 Airbus A320-214, N106US Weehawken, New Jersey. https://www.ntsb.gov/news/events/Pages/Loss_of_Thrust_in_Both_Engines_After_Encountering_a_Flock_of_Birds_and_Subsequent_Ditching_on_the_Hudson_River_US_Airways.aspx (accessed 27 December 2018).

58 E-3 Sentry LX-N90457 Crash (1996). https://commons.wikimedia.org/wiki/File:E-3_Sentry_LX-N90457_Crash,_14_July_1996.jpg (accessed 15 January 2019).

59 Emergency landing after bird strike. Boeing 737-4B6, Amsterdam, Schiphol Airport, 6 June 2010, The Dutch Safety Board, November 2011. https://www.onderzoeksraad.nl/en/media/attachment/2018/7/10/rapport_royal_air_maroc_en_aangepast_web_10012012.pdf (accessed 7 January 2019).

60 Rossiya Ilyushin Il-18 (2006). https://en.wikipedia.org/wiki/Ilyushin_Il-18#/media/File:Rossiya_Ilyushin_Il-18.jpg

61 Antonov_An-124 (2010). https://en.wikipedia.org/wiki/Antonov_An-124_Ruslan#/media/File:Maximus_Air_Cargo_Antonov_An-124-100_Vanzura.jpg (accessed 7 January 2019).

62 https://upload.wikimedia.org/wikipedia/commons/a/aa/Myanmar_Air_Force_Shaanxi_Y-8_MRD.jpg

63 Concorde (2003). https://en.wikipedia.org/wiki/Concorde#/media/File:Air_France_Concorde_Jonsson.jpg

64 https://en.wikipedia.org/wiki/Swearingen_Merlin#/media/File:Fairchild_Swearingen_Merlin.jpg

65 https://en.wikipedia.org/wiki/Antonov_An-8#/media/File:Aeroflot_Antonov_An-8_Osta.jpg

66 1995 Alaska Boeing E-3 Sentry accident. https://en.wikipedia.org/wiki/1995_Alaska_Boeing_E-3_Sentry_accident (accessed 15 January 2019).

67 (2004). *Sharing the Skies: An Aviation Industry Guide to the Management of Wildlife Hazards*, 2e. Transport Canada https://www.ascendxyz.com/wp-content/uploads/regulatory-requirements/Reference%208_Sharing%20the%20Skies.%20An%20Aviation%20Industry%20Guide%20to%20the%20management%20of%20Wildlife%20Hazards.%20Transport%20Canada..pdf (accessed 7 January 2019).

68 https://en.wikipedia.org/wiki/Northrop_T-38_Talon#/media/File:T-38_Talon_over_Edwards_AFB.jpg (accessed 7 January 2019).

69 Bird Strike of a F-111 at Cannon AFB, New Mexico. http://www.usahas.com/AHAS2015/gallery/image.aspx?image=F111%20Bird%20Strike (accessed 7 January 2019).

5

Bird Migration

5.1 Introduction

Bird migration has been known about for more than 3000 years from the writings of Ancient Greeks including Aristotle.

When birds have insufficient resources for their lives, they look for other sources for surviving. Migration is an option for birds that can move away from their current location. Thus, they leave an environment where they would starve and move to habitats that are more hospitable.

Bird Strike in Aviation: Statistics, Analysis and Management, First Edition. Ahmed F. El-Sayed.
© 2019 John Wiley & Sons Ltd. Published 2019 by John Wiley & Sons Ltd.

Migration means two-way journeys – the onward journey from the "home" to the "new" place and the journey back from the "new" place to "home." This movement occurs during a particular period of the year. It is surprising that birds usually follow the same route and cover distances of many kilometers. There is a sort of "internal biological clock" which regulates the phenomenon.

The most common types of migration are those carried out by birds in the spring and the autumn. In the autumn, they travel from breeding grounds in the north to wintering grounds in the south and vice versa in spring (Figure 5.1). Northern areas are characterized by an abundance of food in the summer and little available food in the winter. On the contrary, in southern areas food is available all year around.

Generally, birds are either migratory or sedentary. Birds which migrate are called migratory (Figure 5.2) while birds that remain in one area are called resident or sedentary. Approximately 1800 of the world's 10 000 bird species are long-distance migrants.

Breeding grounds

Autumn migration

Spring migration

Wintering grounds

Figure 5.1 Autumn and spring migration of birds.

Figure 5.2 Bird migration. *Source:* Courtesy USDA-APHIS [20].

(a)

(b)

Figure 5.3 Classification of birds. (a) Migratory (Canada geese). *Source:* Courtesy National Academic Press [28]. (b) Sedentary (sparrow) [29]. *Source:* A male house sparrow in Victoria, Australia in March 2008 by Fir0002, license: Attribution non-commercial Unported 3.0 (GFDL 1.2).

An example of a migratory bird is geese (Figure 5.3a) and an example of a non-migratory or sedentary bird is the sparrow (Figure 5.3b). Large birds such as geese and birds of prey migrate during the day, but small ones like robins (Figure 5.4) migrate during the night.

Migratory birds follow the same migration routes every year. It is interesting to mention that black geese migrate in a V-shaped formation (Figure 5.5).

Figure 5.4 Robin [30]. *Source:* Male Robin Bird, by Dakota Lynch – licensed under: CC BY-SA 3.0.

Figure 5.5 Migration pattern for black geese. *Source:* Courtesy USDA-APHIS [23].

5.2 Why Do Birds Migrate?

For a bird to travel hundreds or thousands of miles between its breeding and non-breeding ranges is a difficult and dangerous journey, one that not all birds survive. So why do birds migrate? There is more than one single reason for different birds to migrate, but it all comes down to survival.

Birds migrate to move from areas of low or decreasing resources (food, water, and nesting locations) to areas of high or increasing resources [1]. Concerning nesting locations, it is noticed that:

- Many birds leave the cold weather grounds in winter for more temperate habitats. Also, birds leave the hot tropical regions in spring to lay eggs in cooler areas in the north in order to provide appropriate weather for raising their delicate chicks.

- Habitats that have abundant food sources all the year round also attract a greater number of predators that can threaten their young in nests. Birds migrate to different habitats to avoid the onslaught of predators. Birds may even migrate to specialized habitats that are nearly inaccessible to predators, such as coastal cliffs or rocky offshore islands.

In the southern hemisphere, birds tend to migrate northward in the spring to nest and take advantage of burgeoning insect populations and budding plants. As winter approaches and the availability of insects and other food drops, the birds move south again. In northern regions, such as Canada or Scandinavia, birds migrate south to escape winter. For temperate regions, such as the UK, about half the species migrate – especially insect-eaters that cannot find enough food during winter. In tropical regions, such as the Amazon rainforest, fewer species migrate, since both weather and food supply are reliable all the year.

5.3   Some Migration Facts

Bird migration has the following nine awesome facts [2]:

1. At least 40% of the world's 10 000 bird species are migratory ones.
2. The highest flying migratory birds are bar-headed geese, which can fly at altitudes of up to 5.5 miles above sea level over the Himalayas in India. However, a Rüppell's griffon vulture has the highest altitude ever recorded when one collided with a plane at an altitude of 7 miles (37 000 ft) in 1975.
3. The Arctic tern has the longest migration route in the world. It can fly more than 49 700 miles in a year.
4. The northern wheatear travels up to 9000 miles each way between the Arctic and Africa, which resembles one of the largest ranges of any songbird.
5. The bar-tailed godwit can fly for nearly 7000 miles without stopping, which is the longest nonstop flight.
6. The fastest bird is the great snipe: It flies at speeds of up to 60 mph for 4200 miles.
7. Many migrating birds die during their trip and so do not return to their starting point. This is due to many reasons like harsh weather, human activities, and collisions with windows, electric cables, wind turbines, etc.
8. Birds bulk up on food in the couple of weeks preceding their migration to store fat which will be used during their long journey. A blackpoll warbler, for example, doubles its weight before flying some 2300 miles without a stop.
9. Some birds, such as the emu (a large Australian flightless bird), do not fly but migrate for miles on foot to find food, while others, such as the penguin, migrate by swimming.

5.4   Basic Types of Migration

There are lots of different kinds of migration. Bird migration can be classified based on length (in time and distance), direction, the reason for migration, and how many of the species migrate. Also, more than one term can be used to describe the migration pattern of a species.

Though not all birds migrate, all species are subject to periodical movements of varying extent. It is noted that birds which live in the northern part of the northern hemisphere have greatest migratory power.

5.4.1 Classification of Migration Based on the Pattern

Migration may be classified into a range of categories [24].

Latitudinal migration: latitudinal migration usually means movement from north to south and vice versa. Most birds live in the land masses of the northern temperate and subarctic zones, where they get facilities for nesting and feeding during summer. They move toward south during winter. As an example, *Sterna paradisaea* (the Arctic tern) breeds in the northern temperate region and migrates to the Antarctic zone along the Atlantic. They cover 22 500 km during this migration.

An opposite, but lesser, movement also occurs in the southern hemisphere. The cuckoo breeds in India and spends the summer in south-east Africa, covering a distance of about 7250 km.

Penguins migrate by swimming and cover a distance of few hundred miles.

Longitudinal migration: longitudinal migration occurs when birds migrate from east to west and vice-versa. An example is the starling (*Sturnus vulgaris*), which is a resident of eastern Europe and western Asia and migrates toward the Atlantic coast. Also, California gulls breed in Utah and migrate westward to winter on the Pacific coast.

Altitudinal migration: altitudinal migration occurs in mountainous regions. Many birds inhabiting the mountain peaks migrate to low lands during winter. Golden plover [24] breed in the Arctic tundra and spend the northern hemisphere winter in the plains of Argentina, covering 11 250 km (Figure 5.6).

Partial migration: partial migration occurs when some individuals of a species migrate while others do not but instead adapt to local conditions. Blue jays of Canada and the northern part of the USA travel southwards to blend with the sedentary populations of the southern states of the USA. Other examples of partial migration are coots and spoonbills. Species such as the red-tailed hawk, herring gull, and golden eagles are partial migrants over much of their North American range [3].

Total (complete) migration: when all the members of the species leave their breeding range during the nonbreeding season it is identified as a total or complete migrant species. Complete migrants travel incredible distances, sometimes more than 15 000 miles (25 000 km) per year. Many North American birds are complete migrants. Most of the complete migrants in North America, Europe, and Asia breed in northern temperate and Arctic areas. The wintering areas for most of the complete North American migrants are South and Central America, the Caribbean basin, and the southernmost United States.

Vagrant or irregular migration: when some of the birds disperse a short or long distance for safety and food it is called vagrant or irregular migration. Herons are an example of birds that show vagrant or irregular migration. Other examples are the black stork, the glossy ibis, the spotted eagle, and the bee-eater.

Daily migration: some birds make a daily journey from their nests, influenced by environmental factors such as temperature, light, and humidity. Examples are crows, herons, and starlings.

Seasonal migration: some birds migrate at different seasons of the year for food or breeding. These are called seasonal migrants. Example are cuckoos, swifts, swallows, etc. They migrate from the south to the north during the summer. These birds are called summer visitors. There are some birds, such as the snow bunting, red wing, shore lark, gray plover, etc., which migrate from north to south during their winter. They are called winter visitors.

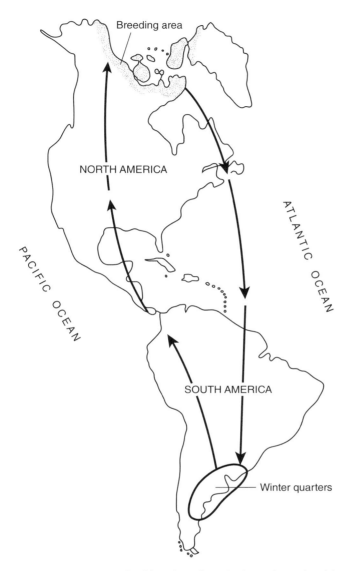

Figure 5.6 Migration of golden plover from Arctic tundra to the plains of Argentina (11 250 km).
Source: Courtesy UNEP [7].

Differential migration: differential migration occurs when all the members of a population migrate, but not necessarily at the same time or for the same distance. The differences are often based on age or sex. For example, as herring gulls (*Larus argentatus*) become older they migrate shorter distances. Male American kestrels (*Falco sparverius*) spend more time at their breeding grounds than do females, and when they migrate, they do not travel as far.

Irruptive migration: irruptive migration occurs in species which migrate in some years and do not in others. Such species migrate if the weather is cold or food is scarce. Examples are rough-legged hawks, snowy owls, crossbills, and redpolls [4].

Loop migration: loop migration is a special migration type, where birds take different routes in their outward and inward routes between their breeding and non-breading areas [5]. A broad range of species from all over the world exhibit loop migration.

5.4.2 Classification of Migration Based on the Type of Motion

Migration may be categorized as flying, walking, or swimming.

Fly (soar) migration: this covers all types mentioned in Section 5.4.1.

Walk migration: some bird species do not fly to their migration destination; instead, they walk. Examples are the ostrich (*Struthio camelus*), the dusky grouse, and the emu (*Dromaius novaehollandiae*). Such species cannot fly and so walk from arid and semi-arid areas to areas having plenty of food and water [6].

Swim migration: most species of penguin (Spheniscidae) migrate by swimming. The routes can cover over 1000 km (620 miles) [6]. Auk species also migrate long distances by swimming [7].

5.4.3 Classification of Migration Based on Distance Traveled

Migration is classified into the following three categories:

- short-distance migration: for example, from higher to lower elevations on a mountainside;
- medium-distance migration: for example, from between states in the USA;
- long-distance migration: for example, from the USA and Canada in the summer to Mexico and further south in the winter.

Examples for short-distance and altitudinal migration are the ruby-throated dipper and cedar waxwing birds (Figure 5.7). They move in response to winter weather.

Examples for long-distance migration are Swainson's thrush, northern pintail, griffon vulture, bar-tailed godwits, and Arctic terns (Figure 5.8). Some bar-tailed godwits have the longest known non-stop flight of any migrant, flying 7000 miles from Alaska to their New Zealand without taking a break for food or drink. Data exists of an Arctic tern that

(a)   (b)

Figure 5.7 Examples for short distance migration birds. (a) Ruby-throated Dipper [31]. (b) Cedar Waxwing [32]. Source: Cedar Waxwing – by Judy Gallagher – licensed under CC BY 2.0.

Figure 5.8 Examples of long-distance birds. (a) Swainson's thrush [33], by Matt Reinbold and Licensed under CC BY-SA 2.0. (b) Northern Pintail [34] by J.M. Garg – Licensed under CC BY-SA 3.0. (c) Griffon vulture [35] in Hai-Bar reserve mount Carmel. Griffon vulture by Pierre Dalous – Licensed under CC BY-SA 3.0. (d) Bar-tailed godwit [36] by Andreas Trepte – licensed under: CC BY-SA 2.5.

left the Farne Island off the British east coast to reach Melbourne, Australia, in just three months. This is a sea journey of over 22 000 km (14 000 miles).

5.4.4   Permanent Residents

Permanent resident birds do not migrate. They can find adequate supplies of food all the year round.

5.5   Flight Speed of Migrating Birds

Most songbirds migrate at about 20–30 mph in still air. Waterfowl and shorebirds can travel at 30–50 mph. Tailwinds or headwinds can drastically influence bird speed and (potentially) their survival.

5.6   Navigation of Migrating Birds

Migrating birds travel thousands of miles in their annual travels. Surprisingly, they travel nearly the same route year after year. First-year birds often make their very first migration on their own. Somehow, they can find their winter home despite never having seen it before, and return the following spring to where they were born.

What is the basis of these navigational skills is not fully known. Birds sense several navigation methods. They can get compass information from the Sun, the stars, and by sensing the Earth's magnetic field. They get information from their position from land-marks seen during the day. Also, smell plays an additional role, at least for homing pigeons.

Some bird species (waterfowl and cranes are examples) follow preferred pathways that include stopover locations rich with food supplies along their annual migration routes.

Smaller birds tend to migrate in broad fronts across the landscape. They have differ-ent routes in spring and fall depending on the availability of food and good weather.

5.7   Migration Threats

Different birds are subjected to deadly threats during migration.

Exhaustion: birds that fly hundreds of miles without rest become exhausted and less wary of potential threats. Thus, they are more likely to collide with obstacles especially if they encounter storms or unfavorable wind.

Starvation: inadequate food supplies cause starvation among migrating birds every year. This is attributed to either habitat destruction, which effectively strands migrating birds without food along with their route, or due to greater feeding competition among large flocks of migratory birds.

Collisions: tens of thousands of migrating birds collide with obstacles such as tall glass buildings, electrical wires, and poles and wind turbines during both spring and fall migrations. These collisions cause fatal injuries.

Predators: predators (like outdoor and feral cats) kill hundreds of thousands of birds each year, including during migration. Migrating birds may be unaware of local preda-tors at stopovers during their journey.

Disease: many birds migrate in large flocks. Consequently, if any are infected, the disease may spread. The overall population may be severely affected.

Pollution: lead poisoning, oil spills, and other pollution sources are harmful to migra-tory birds. Pollution may reduce the amount of food available and make it more difficult for birds to complete their migration successfully.

Natural disasters: natural disasters including hurricanes, blizzards, and wildfires can destroy crucial stopover locations and food sources that birds need to refuel along their journeys.

Hunting: many hunting seasons coincide with migration periods, which makes it more threatening for birds. Illegal hunting is a threat at any time.

Inexperience: the inexperience of birds with migration can be a great threat to their survival. Many young birds travel without guidance from adults. They may not be able to reach their destination if they are unsuccessful in finding adequate food or miss their migration routes.

5.8   Migratory Bird Flyways

5.8.1   Introduction

A flyway is defined as the total geographic area used by a bird species or group of spe-cies throughout its annual cycle [5].

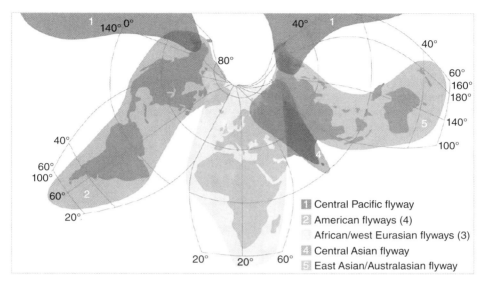

Figure 5.9 Five major flyways for shorebirds. *Source:* Courtesy UNEP [7].

The International Wader Study Group [5] recognized five major flyway groupings for migratory shorebirds (Figure 5.9). The world map can be divided into five flyway areas for shorebirds, with some overlapping at their margins.

1. The Central Pacific Flyway: this extends over the Pacific Ocean from Alaska and far eastern Russia to New Zealand. Though it is traveled by a relatively small number of species, it shows some of the most amazing migrations on Earth.
2. The Americas: this includes North, Central, and South America and the Caribbean as well as the four traditional North American flyways (Pacific, Central, Mississippi, Atlantic).
3. The African Eurasian Migratory Waterbird Agreement (AEWA) area: this includes north-eastern Canada, Greenland (Europe), western Siberia, the Western Central Asian Republics, the Caucasus (Asia), the Middle East, the Arabian Peninsula (Africa), and Madagascar and its associated islands.
4. The Central Asian Flyway (CAF) area: this includes central Siberia, Mongolia, the Central Asian Republics, Iran and Afghanistan, the Gulf States and Oman, the Indian subcontinent, and the Maldives.
5. The East Asian Australasian Flyway (EAAF) area: this includes eastern Siberia, Alaska, Mongolia, Korea, Japan, China, Eastern India, Bangladesh, south-eastern Asia, the Sunda Islands, the Philippines, New Guinea, and Australia (New Zealand is often included).

Several references [8] indicate only three or four major bird migration flyways, as indicated in Figure 5.10. The four regional aggregations are:

- America;
- Africa–Eurasia;
- Central Asia;
- East Asia–Australasia.

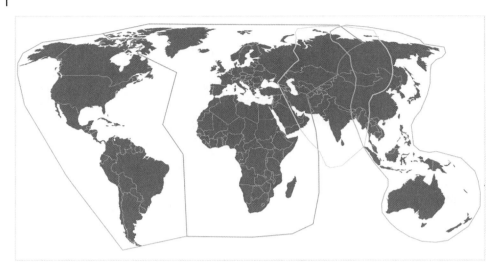

Figure 5.10 Four major bird migration zones. *Source:* Courtesy UNEP [7].

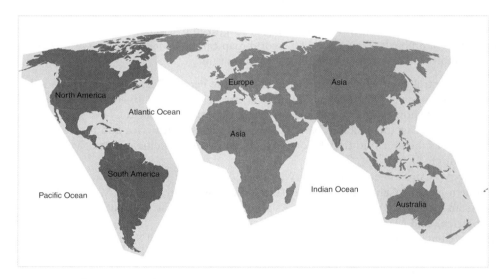

Figure 5.11 Three major bird migration zones. *Source:* Courtesy UNEP [7].

The latter two are sometimes combined as Asia–Pacific, as in Figure 5.11. Thus, these routes are now defined as America, Africa–Eurasia, and Asia–Pacific.

These three regional aggregations may be further described as follows:

Americas Flyway: breeding area in North America and wintering area in the Caribbean and Central and South America.

Africa–Eurasia Flyway: breeding area in Europe and northern Asia and the wintering area in Africa. However, there are vital stopover sites in the Middle East and the Mediterranean.

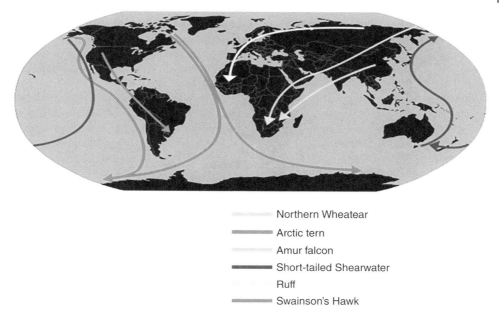

Northern Wheatear
Arctic tern
Amur falcon
Short-tailed Shearwater
Ruff
Swainson's Hawk

Figure 5.12 Migration routes for six birds [11].

East Asia–Australasia Flyway: breeding area in north-east Asia and wintering area in south-east Asia and Australia, with vital stopover sites in China and the Korean Peninsula.

Figure 5.12 illustrates the migration routes for some six birds [11]. The blue route describes the migration route for northern wheatear birds while the red routes illustrate the Arctic tern. The green route illustrates the migration flyway of Amur falcon. Other colors illustrate the migration routes for the short-tailed shearwater, the ruff and Swainson's hawk.

5.8.2 North American Migration Flyways – The Four Ways

The paths followed by migratory birds in North America can be grouped into four general flyways [21]. These are the Atlantic Flyway, the Mississippi Flyway, the Central Flyway, and the Pacific Flyway (Figure 5.13).

These flyways follow the topographical features of the USA in a north to south direction. The main flyway routes depend on the starting place and the final destination. They also match with the abundance of food, shelter, and water. Figure 5.14 shows some of the migrating birds that use the North American migration routes.

5.8.2.1 The Atlantic Flyway
The Atlantic Flyway follows the eastern coast of the USA from the Bahamas to the tip of Maine, with major migration routes intersecting in southern South Carolina and the Delaware Bay.

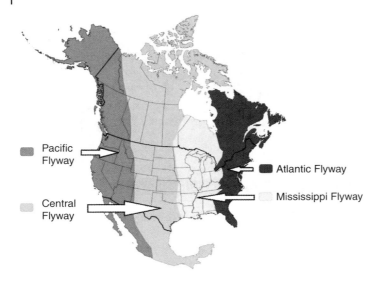

Figure 5.13 North American migration flyways. *Source:* Courtesy U.S. Fish and Wildlife Service [25].

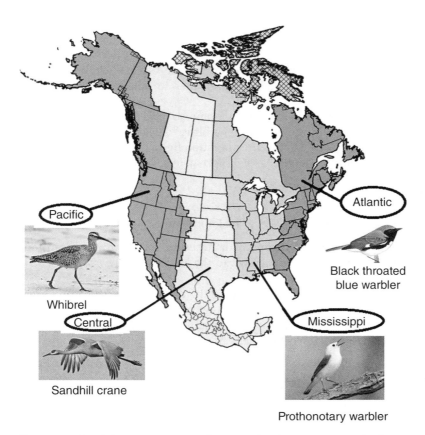

Figure 5.14 North American migration flyways and its famous migrating birds.

The following facts have been identified by migration societies and researchers:

a. Eastern states, namely Alabama, Connecticut, Delaware, Florida, Georgia, Kentucky, Louisiana, Maine, Maryland, Massachusetts, Michigan, Mississippi, New Hampshire, New Jersey, New York, North Carolina, Ohio, Pennsylvania, Rhode Island, South Carolina, Tennessee, Vermont, Virginia, and West Virginia, are covered by the Atlantic flyway [9].
b. About 500 bird species use the Atlantic Flyway.
c. A great portion of the flyway is over or very close to water, including the eastern Atlantic and the Gulf of Mexico.
d. Some species do not travel along the entire route. Some merely migrate a few hundred miles or even over a mountain range that offers a more hospitable climate.
e. The most famous bird species using this flyway are plenty of feeder birds, including American tree sparrow, white-throated sparrow, chipping sparrow, field sparrow, song sparrow, Baltimore oriole, orchard oriole, black-throated blue warbler, chickadee, blue jay, brown thrasher, common redpoll, dark-eyed junco, eastern bluebird, evening grosbeak, hermit thrush, house finch, northern cardinal, northern flicker, pine grosbeak, pine siskin, pine warbler, purple finch, red-winged blackbird, ruby-crowned kinglet, red-breasted nuthatch, and yellow-rumped warbler.
f. A significant section of Canada is also included in the flyway, including Manitoba, Newfoundland and Labrador, New Brunswick, Nova Scotia, Nunavut, Ontario, Quebec, Saskatchewan, Northwest Territories, and Yukon Territory.

5.8.2.2 The Mississippi Flyway

The Mississippi Flyway has the following features:

a. Birds migrate through the middle of North America, tracing the path of the Mississippi river and its tributaries in the US, up to Canada.
b. The states covered by the Mississippi Flyway include Alabama, Alaska, Arkansas, Indiana, Illinois, Iowa, Kansas, Kentucky, Louisiana, Ohio, Oklahoma, Michigan, Missouri, Mississippi, Nebraska, North Dakota, South Dakota, Tennessee, Texas, and Wisconsin. Moreover, this flyway includes the following Canadian provinces and territories: Alberta, British Columbia, Manitoba, Northwest.
c. The migrating species include feeder birds such as the American tree sparrow, chipping sparrow, field sparrow, black-capped chickadee, blue jay, common redpoll, downy woodpecker, evening grosbeak, house finch, indigo bunting, northern cardinal, orchard oriole, purple finch, prothonotary warbler (Figure 5.14), red-headed woodpecker, red-winged blackbird, red-breasted nuthatch, tufted titmouse, and white-breasted nuthatch [10].

5.8.2.3 The Central Flyway

The main route of the Central Flyway has the following features:

a. Its main route follows the east side of the Rocky Mountains from Canada to Southern Colorado, where it moves across Oklahoma and Eastern Texas to continue down the Texas Gulf coast and into Mexico.
b. A second but smaller artery follows the west side of the Rockies and intersects with the main route in Northern Texas.
c. Among the birds migrating this flyway is the sandhill crane (Figure 5.14) [10].

5.8.2.4 The Pacific Flyway

This route has the following features:

a. It hugs the Pacific coast, with a minor migration route right beside the western artery of the Central Flyway.
b. It passes through northern Oregon, central California, and intersects the California–Mexico border.
c. Famous birds for this flyway are gulls, ducks, whimbrel (Figure 5.14), and other water birds.

5.8.3 The Americas Bird Migration

This route has two sub-routes, as described below.

5.8.3.1 North–South Americas

Each autumn (fall), some 3 billion to 5 billion birds return to their wintering grounds in the south after spending the breeding season in the north [16, 17]. Several routes may be identified (Figure 5.15) as follows:

a. Bird species passing through or around the USA include waterfowl, sparrows, finches, waxwings, warblers, sandpipers, cranes, hawks, and eagles[12].
b. Numerous shorebird species reside in the shores and waterways of Central America, the Caribbean, and South America.
c. Many of the Northern forest's vast collection of warblers and other songbirds find inland habitat throughout Central and Northern South America.
d. Some birds, including lesser yellowlegs, blackpoll warbler, and solitary sandpiper species, winter in the Amazon rainforest.

Table 5.1 lists the 1.15 billion birds migrating from the USA to four countries in South America; namely Mexico, Brazil, Columbia, and Venezuela.

5.8.3.2 Alaska's Flyways

Birds come from all around the world to nest in Alaska during the summer, lured by high-quality habitat with plentiful food (Figure 5.16). Among the birds that travel from Alaska to different places [13] are the following:

- to the Pacific: Pacific golden-plover;
- to India: bluethroat;
- to Africa: northern wheatear;
- to South America: American golden plover, pectoral sandpiper, blackpoll warbler;
- to the lower 48 states and the Gulf Coast: American robin, peregrine falcon, Canada goose;
- to the Antarctic: Arctic tern;
- to East Asia/Australia: eastern yellow wagtail, bar-tailed godwit [36], wandering tattler.

5.8.4 Africa Eurasia Flyways

Nearly 4.5 billion birds composed of nearly 185 species, fly north to south and vice-versa between Europe, Asia, and Africa each year. These migration routes pass through

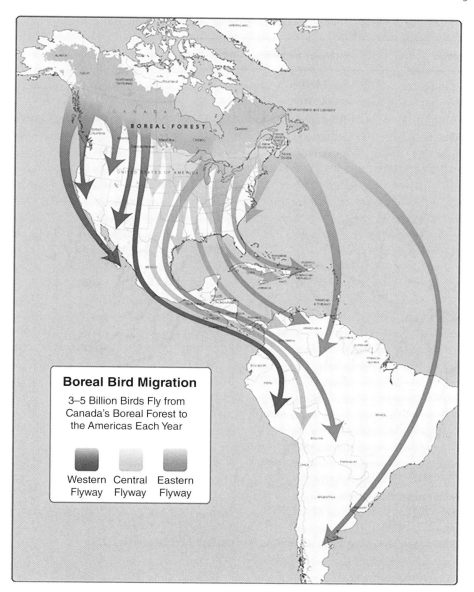

Figure 5.15 Migration routes in the Americas. *Source:* Reproduced with permission of Boreal Songbird Initiative.

Sinai (red color in Figure 5.17), Italy (green color in Figure 5.17), and the Iberian Peninsula (purple color in Figure 5.17). Of the migration routes into Africa, the one through the Middle East appears to be the busiest. Thus, a billion birds, exhausted from over-desert flights between Africa and Europe or Asia refuel in the marshlands and grasslands of the Dead Sea Valley near the Red Sea. Consequently, more than half the aircraft accidents in this area are caused by birds.

Table 5.1 Number of migrating birds through different countries in the Americas.

Destination	Estimated number of wintering birds
USA	1 150 000 000
Mexico	680 000 000
Brazil	200 000 000
Columbia	110 000 000
Venezuela	60 000 000

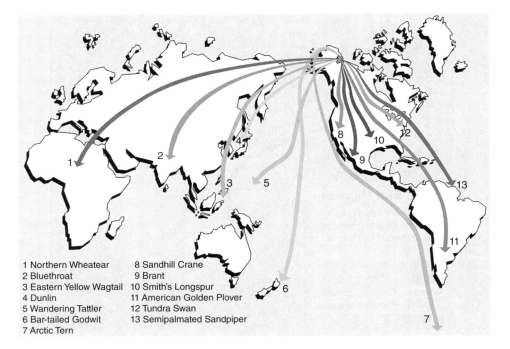

1 Northern Wheatear
2 Bluethroat
3 Eastern Yellow Wagtail
4 Dunlin
5 Wandering Tattler
6 Bar-tailed Godwit
7 Arctic Tern
8 Sandhill Crane
9 Brant
10 Smith's Longspur
11 American Golden Plover
12 Tundra Swan
13 Semipalmated Sandpiper

Figure 5.16 Alaska's migratory birds. *Source:* Courtesy U.S. Fish & Wildlife Service [22].

In the interior of Africa and around its coastal areas bird migration is irregular, ranging from local to international and often driven by climatic factors.

Western and eastern Palearctic–African migration routes comprise large numbers of birds traveling between Europe and Africa twice a year (Figure 5.18). Approximately 2.1 billion songbirds and near-passerine birds migrate from Europe to Africa in the autumn [12, 26].

Many northern and eastern European bird species have pronounced migratory tendencies while many species from western Europe are more sedentary. Some birds in Europe spend the colder months in the southwestern part of the continent or the Mediterranean region while others migrate to Africa south of the Sahara (Figure 5.19).

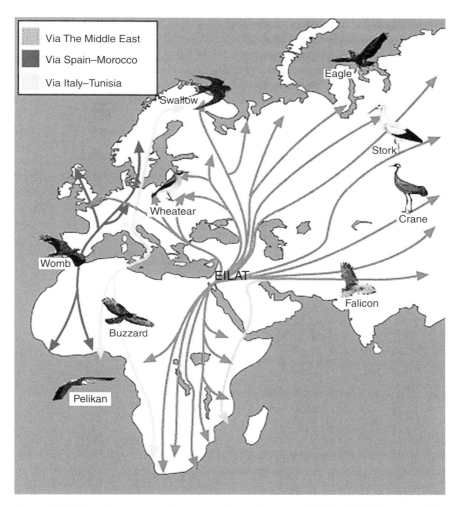

Figure 5.17 Africa Eurasia migration routes. *Source:* Courtesy U.S. Fish & Wildlife Service [12].

Geographical terrains determine several main routes. For example, the Alps are an important barrier to migratory birds. About 150 species travel westward and south-westward while others travel southeastward.

Some details of bird species and migration routes are described here:

- Tits, goldfinches, and blackbirds in northern Europe have short migration flights.
- Large flocks of starlings pass the winter in North Africa.
- Warblers, flycatchers, and wagtails use different routes to cross the Mediterranean. Next, they follow one of the following three routes, either to Sierra Leone on the west coast, Tanzania on the east coast, or straight ahead to the tip of the continent.
- Golden orioles and red-backed shrikes go to East Africa via either Greece or Egypt.
- Swallows spend the winter in the Congo River region, South Africa, and some coastal areas of West Africa.

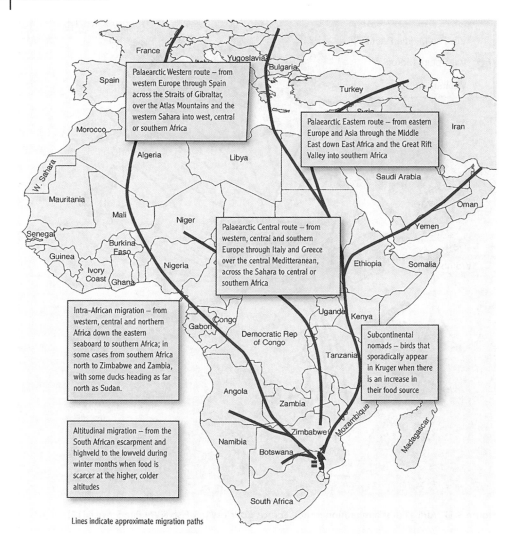

Figure 5.18 Bird migration routes in Africa. *Source:* Reproduced with permission of Siyabona Africa (Pty)Ltd. [37].

- The stork migrates along two well-defined flyways. Storks in the eastern part migrate via a route over the straits of the Bosporus, through Turkey and Israel, to East Africa. Storks in the western part fly southwestward through France and Spain, past the Strait of Gibraltar, and reach Africa by way of West Africa.
- Ducks, geese, and swans winter partly in tropical Africa and partly in western Europe. Those wintering in Africa stay in the lake and river regions between Senegal in western Africa and Sudan in eastern Africa.
- Most shorebirds nest in flat, treeless regions of the Arctic and winter along the seacoasts from western Europe to South Africa.

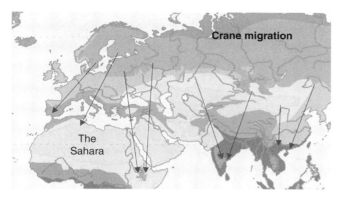

Figure 5.19 Migration routes of European bird species. *Source:* Courtesy UNEP [7].

Figure 5.20 East Asian–Australian migration zone. *Source:* Courtesy UNEP [5].

5.8.5 East Asian–Australian Flyways

The East Asian–Australasian Flyway (EAAF) extends from within the Arctic Circle in Russia and Alaska, through both East and South-East Asia, in a southern direction, to Australia and New Zealand. It encloses 22 countries, as displayed in Figures 5.20 and 5.21. Australia regularly holds 36 species of migratory shorebirds during the northern winter. It includes two stop sites each of which supports up to 3 million birds.

Figure 5.21 illustrates the three flyways of the Asia–Pacific zone, which overlap in breeding and staging areas. Thirty-four species of the shorebirds in Australia are migratory while 15 others are resident.

Most of the migrant species breed in northern China, Mongolia, Siberia, and Alaska during June and July, and then migrate to Australia [15]. For the non-breeding season, more than 1 million shorebirds migrate to and from Australia every year. Several millions other worldwide travel a great distance between their breeding and non-breeding habitats.

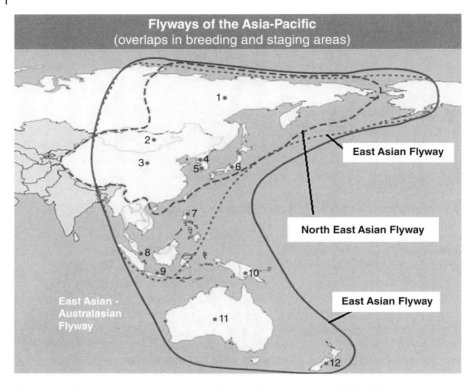

Figure 5.21 Three East Asian–Australasian flyways. *Source:* Courtesy UNEP [27].

The migration routes [14] for one of the shorebird species, the crane family, are shown in Figure 5.22.

5.9 Radio Telemetry

Radio telemetry is a technique used to determine the movements of birds over areas ranging from small areas, such as breeding territories of resident bird species, to the large areas covered by international migratory species [18].

Telemetry techniques (including satellite ones) have long been used in the study of bird migration routes, nocturnal migrations, flight altitudes, bird numbers, and daily movement patterns. A significant database exists – a compilation of historical information on movement patterns from many areas of the world [19].

The idea of radio telemetry is to attach a radio transmitter to a bird and track the signal to determine its movement. Thus, telemetry can provide a history of detailed bird movements. Recent radio transmitters are either a platform terminal transmitter (PTT) or global positioning system (GPS) type, with capabilities far beyond the old conventional VHF radio transmitters.

All transmitter types (PTT, GPS, and VHF) operate using the same basic principles of emitting an electromagnetic signal at a specified frequency which is detected by

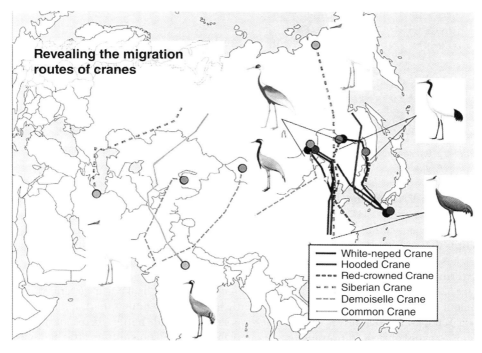

Figure 5.22 The migration routes for cranes. *Source:* Courtesy UNEP [27].

receivers tuned to this frequency. Advanced transmitters use orbiting satellites to receive and relay transmitter signals. Table 5.2 lists a comparison between the three types of radio transmitters.

Table 5.2 A comparison between the three types of radio transmitters [18].

	Types of radio transmitters		
	VHF	**PTT**	**GPS**
Transmitter wt.	<1 g to 12 g	12–18 g	30–60 g
Species	>20 g	>500 g	>1 kg
Minimum cost	US$ 100	US$ 3200	US$ 3800
Attachment	Anchor, feather, implant	Collar, backpack, implant	Collar, backpack, implant
Power source	Battery	Battery or solar	Battery or solar
Duration	Day to months	Months to years	Months to years
Range	0.1 to 100+ km	Unlimited	Unlimited
Tracking	Manual	Satellite	Satellite
Tracking interval	Continuous	4 h	Continuous
Accuracy	±5 m to 1 km	±100–200 m	±10–20 m
Frequency	VHF	UHF	UHF

Figure 5.23 Backpack harness for attaching GPS transmitter. *Source:* Courtesy Roy Dennis Wildlife Foundation [38].

Employing VHF telemetry for tracing long-range migratory species has many logistic difficulties because VHF telemetry requires the mobilization of telemetry tracking teams across enormous areas, some of which may be inaccessible to observers on the ground. VHF telemetry studies require considerably more logistical planning than satellite-based studies, mainly due to the manual tracking efforts that are required.

External attachment of transmitters is via a neck collar, backpack harness, or leg. Figure 5.23 illustrates a backpack collar type. Neck collars and backpacks generally have excellent transmitter retention and are currently the only attachment methods available for PTT and GPS transmitters. Harness designs should provide better fits to a particular species to avoid abrasions or impede wing movement. VHF transmitters attached with leg bands also have excellent retention, but problems with transmission range have been noted, possibly due to the shorter antennae and proximity to the ground.

References

1 The Cornell Lab of Ornithology (2007). The Basics of Bird Migration: How, Why, And Where. https://www.allaboutbirds.org/the-basics-how-why-and-where-of-bird-migration (accessed 8 January 2019).

2 Lockhart, J. (2012). 9 Awesome Facts About Bird Migration. http://www.audubon.org/news/9-awesome-facts-about-bird-migration (accessed 8 January 2019).

3 Hochachka, W.M., Wells, J.V., Rosenberg, K.V. et al. (1999). Irruptive migration of common redpolls. *The Condor* 101 (2): 195–204. https://doi.org/10.2307/1369983.

4 Kirby, J.S., Stattersfield, A.J., Butchart, S.H.M. et al. (2008). Key conservation issues for migratory land- and waterbird species on the world's major flyways. *Bird Conservation International* 18: S74–S90.

5 Lambertini, M. (2012). *A Bird's Eye View on Flyways – A Brief Tour by the Convention on the Conservation of Migratory, Species of Wild Animals*, **2**e. UNEP/CMS.

6 Elphick, J. (2007). *The Atlas of Bird Migration: Tracing the Great Journeys of the World's Birds*. London: Natural History Museum.

7 Galbraith, C.A., Jones, T., Kirby, J. and Mundkur, T., A REVIEW OF MIGRATORY BIRD FLYWAYS AND PRIORITIES FOR MANAGEMENT, CMS Technical Series Publication No. 27, UNEP/CMS 2014.

8 Bird Migration: Birds of the Atlantic Flyway. http://www.birdfeeders.com/blog/wild-bird/atlantic-flyway-migration (accessed 8 January 2019).

9 Flyways of the Americas. http://www.audubon.org/birds/flyways (accessed 8 January 2019).

10 Bird Migration: Birds of The Mississippi Flyway. https://www.perkypet.com/articles/mississippi-flyway-migration (accessed 15 January 2019).

11 Bird migration. https://en.wikipedia.org/wiki/Bird_migration (accessed 8 January 2019).

12 Alaska Migratory Bird Calendar Contest 2016, Working Together to Save Migratory Birds. https://www.fws.gov/uploadedFiles/Region_7/NWRS/Zone_2/Selawik/PDF/Bird%20Calendar%20Contest%20packet_2016.pdf (accessed 8 January 2019).

13 Zimmer, C. (2017). On Long Migrations, Birds Chase an Eternal Spring. https://www.nytimes.com/2017/01/05/science/on-long-migrations-birds-chase-an-eternal-spring.html?rref=collection%2Fsectioncollection%2Fscience&action=click&content Collection=science&region=rank&module=package&version=highlights&content Placement=1&pgtype=sectionfront&_r=0 (accessed 15 January 2019).

14 Ramachandra, T.V., Mahapatra, D.M., Boominathan, M. K. et al. (2011). Environmental Impact Assessment Of The National Large Solar Telescope Project And Its Ecological Impact In Merak Area, CES Technical Report XXM.

15 East Asia/ Australasia Flyway. http://www.birdlife.org/sites/default/files/attachments/8_East_Asia_Australasia_Factsheet.pdf (accessed 8 January 2019).

16 Buhnerkempe, M.G., Webb, C.T., Merton, A.A. et al. (2016). Identification of migratory bird flyways in North America using community detection on biological networks. *Ecological Applications* 26 (3): 740–751.

17 Saha, P. (2016). See the Millions of Places Migrating Birds Have Gone - in One Gif, A new animation from the Cornell Lab of Ornithology tracks 118 species along their journeys across the Americas. *Audubon* 27: 2016. https://www.audubon.org/news/see-millions-places-migrating-birds-have-gone-one-gif (accessed 15 January 2019).

18 Fuller, M.R., Millspaugh, J.J., Church, K.E., and Kenward, R.E. (2005). Wildlife radiotelemetry. In: *Techniques for Wildlife Investigations and Management* (ed. C.E. Braun), 377–417. Bethesda, USA: The Wildlife Society.

19 Whitworth, D., Newman, S., Mundkur, T. and Harris, P. (2007). Wild Birds and Avian Influenza: An Introduction to Applied Field Research and Disease Sampling Techniques, FAO Animal Production and Health Manual No. 5. http://www.fao.org/docrep/010/a1521e/a1521e00.htm (accessed 15 January 2019).

20 Moseley, B. Migratory Bird Habitat Initiative: a success for Chambers County landowner. https://www.nrcs.USDA-APHIS.gov/wps/portal/nrcs/detail/tx/home/?cid=nrcs144p2_003217 (accessed 8 January 2019).

21 Arctic National Wildlife Refuge: Bird Migration Routes, U.S. Fish & Wildlife Refuge. https://www.fws.gov/uploadedFiles/Region_7/NWRS/Zone_1/Arctic/PDF/birdmigration.pdf (accessed 8 January 2019).

22 U.S. Fish & Wildlife Service (2018). Birds Keep Us Connected to the Arctic https://medium.com/usfws/birds-keep-us-connected-to-the-arctic-cd0d1c099f12 (accessed 8 January 2019).

23 Dolbeer, R.A., Wright, S.E., Weller, J. and Begier, M.J. (2012). Wildlife Strikes to Civil Aircraft in the United States, 1990–2010. Serial Report Number 17. FAA and USDA-APHIS.

24 Tanika, M., Bird Migration: Definition, Types, Causes and Guiding Mechanisms. http://www.biologydiscussion.com/zoology/birds/bird-migration-definition-types-causes-and-guiding-mechanisms/41286 (accessed 8 January 2019).

25 Flyways, U.S. Fish & Wildlife Service. https://www.fws.gov/birds/management/flyways. php (accessed 15 January 2019).

26 Kruger Park Birding: Bird Migration Routes (2017). http://birding.krugerpark.co.za/ birding-in-kruger-migration-routes.html (accessed 8 January 2019).

27 Hagemeijer, W. and Mundkur, T., Migratory flyways in Europe, Africa and Asia and the spread of HPAI H5N1, UNEP/CMS 2006.

28 DeFusco, R.P. and Unangst, E.T. (2013). Airport Wildlife Population Management: A Synthesis of Airport Practice. Airport Cooperative Research Program, Synthesis 39. https://www.nap.edu/read/22599/chapter/1 (accessed 29 December 2018).

29 Sparrow (2008). https://en.wikipedia.org/wiki/Sparrow#/media/File:House_Sparrow_ mar08.jpg (accessed 8 January 2019).

30 American robin (2013). https://en.wikipedia.org/wiki/American_robin#/media/ File:American_Robin_Close-Up.JPG (accessed 8 January 2019).

31 Ruby-throated hummingbird (2003). https://en.wikipedia.org/wiki/Ruby-throated_ hummingbird#/media/File:Rubythroathummer65.jpg (accessed 9 January 2019).

32 Cedarwaxwing (2018). https://en.wikipedia.org/wiki/Cedar_waxwing#/media/ File:Cedar_Waxwing_-_Bombycilla_cedrorum,_George_Washington%27s_Birthplace_ National_Monument,_Colonial_Beach,_Virginia_(39997434862).jpg (accessed 8 January 2019).

33 Swainson's thrush (2009). https://en.wikipedia.org/wiki/Swainson%27s_thrush#/media/ File:Catharus_ustulatus_-North_Dakota-8a.jpg (accessed 8 January 2019).

34 Northern pintail (2007). https://en.wikipedia.org/wiki/Northern_pintail#/media/ File:Northern_Pintails_(Male_%26_Female)_I_IMG_0911.jpg (accessed 8 January 2019).

35 Griffon vulture (2013). https://en.wikipedia.org/wiki/Griffon_vulture#/media/ File:Gypful.jpg (accessed 8 January 2019).

36 Bar-tailed godwit (2011). https://upload.wikimedia.org/wikipedia/commons/c/ca/ Bar-tailed_Godwit.jpg (accessed 8 January 2019).

37 http://birding.krugerpark.co.za/images/kruger-bird-migration-map.jpg

38 http://www.roydennis.org/o/wp-content/uploads/2011/10/CULLEN34.jpg

6

Bird Strike Management

Bird Strike in Aviation: Statistics, Analysis and Management, First Edition. Ahmed F. El-Sayed.
© 2019 John Wiley & Sons Ltd. Published 2019 by John Wiley & Sons Ltd.

6.1 Introduction

Throughout history, humans have been intrigued and inspired by the beauty of birds and their ability to fly. However, when birds and aircraft started to share the sky, bird–aircraft strikes occurred. Bird strikes are increasing day by day due to the following reasons:

1) Increased air traffic: as an example, at any instant, there are 4000–6000 commercial aircraft flying in the USA and some 50000–60000 daily commercial flights. There are 13000–16000 commercial aircraft all over the world, at any time. Moreover, there is a sharp increase in military aircraft movements due to its vital role in logistical and tactical military operations [1, 2].

2) There have been drastic increases in the population of bird species commonly involved in bird strikes in recent decades, especially large-bodied species like cranes, gulls, herons, pelicans, and raptors. They have been even accustomed to living in or near airports. As an example, Canadian geese numbers have increased from 1 million to 4 million in the period 1980–2007, in the US and Canada.

3) Commercial jets today have three particular features – namely they are faster, are much quieter, and have only two engines. Birds either cannot hear or avoid the oncoming aircraft. Moreover, if birds are ingested simultaneously in both engines of a twin-engine aircraft, such an aircraft is more likely to have difficulties than a three- or four-engine aircraft.

4) Airport managers face increased concerns about airport liability in the aftermath of damaging bird strikes.
   Figures 6.1 and 6.2 show hazardous events of bird strike with an aircraft and a helicopter.

6.2 Why Birds Are Attracted to Airports

Birds are attracted to airports since they provide food, habitat, shelter, water, and a secure environment [3, 32].

6.2.1 Food

Food for birds is available at airports. These are some of the reasons:

- Small birds and rodents/other animals are harbored by tall and poorly maintained grass stands and borders (Figure 6.3).
- Careless waste disposal practices by some staff in restaurants and airline flight kitchens (Figure 6.4).
- Nearby landfill sites or sewage outlets.

Figure 6.1 Bird strike with an aircraft close to an airport. *Source:* courtesy USDA-Aphis [1].

Figure 6.2 Hazard of bird strike with a helicopter [56].

6.2.2 Water

All birds are attracted to open water for its different roles in avian life, including drinking, feeding, bathing, and roosting. Rainy periods may result in temporary water pools at airports. Many airports have permanent bodies of water near or between runways for landscaping, wastewater purposes, or flood control (Figure 6.5). Permanent sources of water provide plenty of food for birds, including invertebrates, small fish, insect larvae, tadpoles, frogs, and edible aquatic plants.

Figure 6.3 Grass attractants food for birds. *Source:* courtesy Transport Canada [32].

Figure 6.4 Refuse as food for birds. *Source:* courtesy Transport Canada [32].

6.2.3 Cover

Birds need cover for resting, roosting, and nesting. These requirements are provided by trees, shrubs, patches of weeds, and airport structures (Figure 6.6). Generally, an area that is free from human disturbance provides a suitable resting/roosting site for some bird species.

Figure 6.5 Rain as a water source for birds. *Source:* courtesy USDA-APHIS [1].

Figure 6.6 Cover for birds. *Source:* courtesy USDA-APHIS [1].

6.3 Misconceptions or Myths

It is important to understand the misconceptions in bird strike management. Table 6.1 looks at the misconceptions or myths associated with bird strikes and gives the correct information.

6.4 The FAA National Wildlife Strike Database for Civil Aviation

We must understand the various aspects of the bird strike problem before it can be solved.

Table 6.1 Misconceptions or myths associated with bird strike [4–8].

Myth number	Statement	Correction
1	Birds do not fly at night	Birds fly at night and have struck thousands of aircraft (Figure 3.35)
2	Birds do not fly in poor visibility environment such as fog, cloud, rain, or snow	Birds can fly and strike aircraft in all weather conditions
3	Birds can detect weather radar and airplane landing lights, thus avoid the airplane	Generally, no. Even if so, birds cannot avoid collisions with aircraft
4	Jet engine spinner markings help to repel birds	Most catastrophic aircraft accidents are due to bird ingestion into engines
5	Birds try to avoid airplanes due to aircraft and engine noise	Birds became accustomed to aircraft and engine noise and even fly toward the engines, leading to serious damage of the engine core
6	Birds dive to avoid any approaching airplane	Bird strikes in the approach phase of flight are 18% of the total bird strike with different aircraft modules, as demonstrated in Figure 3.24
7	Bird strike cannot cause serious airline accidents	Nine large airliners had major accidents between 1988 and 2012
8	Bird strikes are rare	FAA reported over 169 856 wildlife strikes (with 97% is due to bird strike) in the interval 1990–2015
9	Bird strikes are a new problem, which was not encountered some 20 or 30 years ago	Bird strike hazards are an old problem since the Wright Brothers in 1905.
10	Large aircraft are designed to withstand all bird strikes	Though large commercial aircraft, such as Boeing 747, 757, 767, 777, and 787 and Airbus 320, 330, 340, 350, and 380, are certified to be able to continue flying after impacting a 4-lb bird, they encounter severe damage and even complete destruction with many fatalities and serious injuries due to impact with a single or few large-size birds.
		Even flocks of small birds, such as starlings and blackbirds, and a single medium-sized bird, such as hawks, gulls, and ducks, can cause engine failure which may lead to substantial damage to the engine or airframe.
11	If a bird is ingested into the engine of a transport category airplane during takeoff and the engine has failed, then the airplane will crash	Transport category aircraft powered by two or more engines are designed so that if one engine fails, then the airplane will have enough power from the remaining engine(s) to complete the flight.

Table 6.1 (Continued)

Myth number	Statement	Correction
12	It is impossible to keep birds away from airports	It is possible to move birds away from an airport by one or more of the following methods: • make the airport area and its surroundings unattractive for birds; • scare the birds; • reduce the bird population.
13	It is illegal to kill birds only for the sake of protection aircraft	• Bird species that pose a threat to aircraft and that are not protected by law (such as pigeons and starlings in North America) may be killed. • If non-lethal techniques are not effective in keeping birds away, the airport authority may allow killing a limited number of birds such as ducks, geese, gulls, and herons. • However, it is prohibited to kill endangered bird species under any circumstances.
14	Killing off all birds in the airport will eliminate the bird strike problem	• Killing off all birds at an airport (if legal) will not solve the problem. • Since the airport is an integral part of the local ecosystem (including all bird, plant, or animal species), then eliminating any one problem species will only lead to other species taking its place. • A long-term solution to this problem necessitates a combination of bird control measures and habitat management.
15	Except for the very rare accident, bird strikes are only a nuisance to airline operators.	• Bird strike may lead to both human loss and injury as well as annual losses of more than US$ 1.2 billion for airlines worldwide. • Even minor damage to an airliner due to bird strike is usually not covered by aircraft hull or engine insurance. Thus it directly affects airline profits. • Damage to any aircraft module due to bird strike needs either repair or replacement, which has direct costs (including material and labor) as well as indirect costs (including changing aircraft and keeping the aircraft out of revenue service).
16	Bird strikes influence only flying aircraft	• Bird strikes are influenced by many aviation, social, and policy issues. • New restrictions for usage of lands (say within 5 miles from airports) as landfill or wildlife refuges become an urgent necessity since they attract hazardous wildlife. • Bird populations are affected by past and present wildlife policies and habitat management. • If bird strikes led to aircraft accidents, families and friends of victims are affected. • If bird strikes led to fuel dumping (which may be tens of tons) for a safe landing, it might have environmental consequences.

Firstly, we must collect all data available on bird collisions with aircraft; however, much data is missing. The FAA Wildlife Strike Database contains only records of reported wildlife strikes since 1990.

The FAA database is constructed through accurate reporting of wildlife strikes, though the focus here is on bird strikes. Form 5200-7, *Bird/Other Wildlife Strike Report*, is the standard form for the voluntary reporting of bird and other wildlife strikes with aircraft. Strikes can be reported online.

The FAA database has the following features:

- it represents only the information received from airlines, airports, pilots, and other sources;
- it may not contain all bird strike incidents;
- it may contain no incidents for a certain airport;
- the most highly reported area for bird strikes includes the states of New York, New Jersey, and Florida.

The FAA database provides:

- a scientific basis for identifying risk factors;
- the corrective actions needed at airports;
- the effectiveness of these corrective actions;
- identification for wildlife-resistant aircraft and engines;
- a means for engineers, biologists, and safety analysts to better understand national and regional trends in strikes.

Wildlife strike reporting is voluntarily and represents information received from airports, airlines, pilots, and other sources [9].

FAA database includes the following three sections:

- Search the Database [10];
- Report a Strike;
- Edit a Strike Report.

The FAA database is operated and maintained by Embry-Riddle Aeronautical University in Prescott, Arizona, USA.

The database includes the following Microsoft Excel spreadsheets:

- Reported Strike Totals Per Year by Airport: This report shows details of accidents in selected airports and the number of reported strikes at each since 2003. Since bird strike reporting is voluntary, then some airports have complete information while others do not. Results can be downloaded in Excel format. An example for this sheet as displayed in Excel format is shown in Table 6.2.
- Reported Strikes Breakdown by Species: This report classifies some reported strikes per species since 1990. The report also identifies the percentage of strikes caused by multiple birds and the percentage of strikes causing damage to the aircraft. Also, it can be downloaded in Excel format.
- USGS (United States Geological Survey) Banded Birds Involved in Strikes: This report is concerned with birds banded (ringed) by USGS that are then involved a bird strike.
- Strikes at any Airport, FAA Region, or State: Enter information such as airport, state, FAA region, time, or species, to see all the strikes reported that match that criteria. Results can be downloaded in Excel format.

Table 6.2 Reported strike totals per year by US airport from 2006 to 2016.

Airport ID	Airport	State/FAA Region	2006	2007	2003	2009	2010	2011	2012	2013	2014	2015	2016	Total
KMSP	MINNEAPOLIS-ST PAUL INTL/WOLD-CHAMBERLAIN ARPT	MN/AGL	75	96	86	90	98	97	110	104	127	116	13	1012
KBWI	BALTIMORE/WASH INTL THURGOOD MARSHAL ARPT	MD/AEA	57	85	71	106	106	105	99	99	108	141	14	991
KIAH	GEORGE BUSH INTERCONTINENTAL/HOUSTON ARPT	TX/ASW	32	44	37	137	114	100	97	82	155	164	21	983
KPDX	PORTLAND INTL (OR)	OR/ANM	108	114	113	82	63	72	112	86	108	102	17	977
KBOS	GENERAL EDWARD LAWRENCE LOGAN INTL ARPT	MA/ANE	70	65	81	76	119	101	99	127	118	92	21	969
KHOU	WILLIAM P HOBBY ARPT	TX/ASW	70	66	90	125	104	127	102	85	102	89	8	968
KCLE	CLEVELAND-HOPKINS INTL ARPT	OH/AGL	49	82	138	92	148	96	89	106	75	66	18	959
KLAX	LOS ANGELES INTL	CA/AWP	77	39	44	75	86	110	118	111	100	105	33	898
KIAD	WASHINGTON DULLES INTL ARPT	DC/AEA	87	84	74	72	103	96	71	84	101	97	13	882
KTEB	TETERBORO AIRPORT	NJ/AEA	72	86	87	98	66	76	75	98	100	75	11	844
KIND	INDIANAPOLIS INTL ARPT	IN/AGL	56	60	70	85	79	80	97	69	125	96	16	833
KSEA	SEATTLE-TACOMA INTL	WA/ANM	57	53	59	67	88	121	71	57	89	93	12	767
KSFO	SAN FRANCISCO INTL ARPT	CA/AWP	79	57	54	68	69	80	124	71	70	73	16	761
KMDW	CHICAGO MIDWAY INTL ARPT	IL/AGL	55	41	86	77	70	76	80	78	69	106	22	760
KPHX	PHOENIX SKY HARBOR INTL ARPT	AZ/AWP	93	100	78	77	84	59	49	58	66	72	23	759

- Airport Wildlife Strike Summary and Risk Analysis Reports: These reports provide a summary of strike data for any selected airport. The reports will provide risk analysis for bird species and risk management priorities.

6.5 Management for Fixed-Wing Aircraft

6.5.1 Reduction of Bird Strike Hazard

The bird strike hazard faced by an individual airport depends on many factors [1], including:

- type and volume of air traffic;
- bird migration routes and populations;
- conditions of the local wildlife habitat.

As outlined in Section 6.1, birds are attracted to an airport environment due to the availability of food, water, and habitat. Most of the bird strikes are encountered within the immediate airport for civil aircraft, while for military aircraft bird strikes occur above or very close to air force bases.

Therefore, bird strike management starts by correcting the environment on and near the airport, which is usually not an easy task. Active methods may vary from using predators for repelling wildlife to lethal control of wildlife.

First, it is needed to define when wildlife (bird) strike has occurred. It is confirmed by one of the following:

1) a report by a pilot describing one or more bird strikes;
2) a sudden action taken by pilots to avoid a strike with birds, such as aborted takeoff or landing, high-speed emergency stop, or aircraft leaving pavement area;
3) a report by ground personnel who saw an aircraft strike by one or more birds;
4) an identification by maintenance personnel confirming that aircraft damage was caused by a bird strike;
5) remains of bird found within 200 ft of the centerline of a runway (unless another reason for the bird's death is identified).

The hazard of bird strike can be minimized by using one or more of the following three methods [4]:

- awareness;
- airfield bird control and avoidance;
- aircraft design.

6.5.2 Awareness

Awareness means recognizing the presence, problems, and danger of birds at and around an airport. It does not only apply to birds but also applies to land use and other activities in the vicinity of the airport. Awareness should lead to a careful study of the ecology and behavior of the relevant species, the problems they cause, and possible measures and solutions [11].

Flight crews can reduce the possibility and damage of a bird strike by increased awareness and by following recommended procedures [6].

6.5.3 Airfield Bird Control

Plans for airfield bird control include the following four categories [12]:

- control of airport and surroundings;
- ATS providers;
- air operators;
- pilots.

These categories will be thoroughly described in Sections 6.6–6.9

6.5.4 Aircraft Design

Aircraft manufacturers, such as Boeing, Airbus, Embraer, etc., and engine manufacturers, such as Rolls Royce, GE, Pratt & Whitney, have added new design criteria for bird strike resistance. Moreover, different aviation authorities, such as the FAA, EASA, and Transport Canada Civil Aviation TCCA, have set additional measures for certification of both new airframes and new engines to survive possible bird impacts.

Details of such new constraints will be thoroughly discussed in Section 6.10. In brief, current research and development (R&D) carried out by aircraft and engine manufacturers focuses on the following three areas:

- certification standards;
- design and material selections of modules;
- impact tests.

6.6 Control of Airport and Surroundings

Airport control has both active and passive management programs. Adopting both techniques can successfully minimize the threats from hazardous bird populations. These techniques vary in cost and effectiveness, depending on the situation. Active control is a short-term control which aims at dispersing birds from an airfield. On the contrary, passive control techniques (commonly referred to as the habitat-management elements) have a long-term nature. Passive control techniques aim to eliminate or reduce the attractive conditions for birds [4].

6.7 Active Controls

The presence of birds on runways, taxiways, or infields create a potential hazard, so they must be dispersed before aircraft ground movements or flights. However, birds can move quickly and unpredictably and thus create an immediate hazard. Active control methods are subdivided into portable and static systems [13]. The levels of sophistication, and hence cost, and availability are highly variable.

In summary, bird/wildlife deterrent devices/techniques can be broadly divided into five categories.

- Auditory (or bioacoustic): artificial noises, ultrasonics, and high-intensity sound are either ineffective or unsafe. Complex radio-controlled sound generators resemble static acoustic methods. On the contrary, pyrotechnic pistols and vehicle-mounted distress call apparatus are mobile acoustic methods. Unfortunately, much of the information on acoustic methods is unpublished and not generally available.
- Visual: visual techniques range from extremely effective to ineffective. The simple scarecrow is a simple static visual device, while handheld lasers are mobile visual devices.
- Lethal: lethal techniques are generally untested. However, they are less effective but costly means for population control. In the UK lethal methods are licensed by the Department for Environment, Food and Rural Affairs under the Wildlife and Countryside Act. Traps represent a static lethal method while a shotgun is an example of a mobile lethal method.
- Chemical techniques: these are relatively expensive and are time-consuming and difficult to apply. Few chemicals are licensed for use as bird repellents.
- Exclusion techniques: exclusion techniques are usually extremely effective but expensive. Therefore, they are used for high-value crops or costly damage.

It is important to state that no single method may be employed alone for dispersing birds in all conditions. Using a combination of different dispersal methods provides the best defense against immediate hazards. Moreover, using several techniques together, or in an alternating manner, remain more effective over a longer period. Note that no permits are needed for using non-lethal methods against migratory birds.

6.7.1 Auditory (or Bioacoustic) Methods

6.7.1.1 Pyrotechnics

Pyrotechnics have proved effective in dispersing birds at airports. They are noise-producing devices, commonly used in bird dispersal. However, before they are used, the US Fish and Wildlife Service (USFWS) must be consulted to be sure that the process will not affect any Federal threatened or endangered species. Scare cartridges are a commercially available pyrotechnic, and are fired from a 12-gauge shotgun or an NJ-8 Very pistol. Cartridges projected from a shotgun have ranges of 45–90 m, while pistols have a range of approximately 25 m, and then explode. Bird scaring cartridges can produce noise levels of up to 160 dB at varying ranges. Both the cartridges and the gun require a firearms certificate [14]. As illustrated in Figure 6.7, even showing guns and misfiring are effective ways of scaring birds.

Pyrotechnics can be used not only to disperse flocks of birds but also in moving the birds in a desired direction, which is mostly away from runways. Coordination with both control tower and base security forces must be performed before using pyrotechnics to guarantee that birds will not be directed into the path of arriving or departing aircraft.

Munitions purchased for authorized Bird Aircraft Strike Hazard (BASH) are either centrally managed, which can be ordered only through the base supply system, or are commercial off-the-shelf (COTS) which can be purchased only using the government purchase card (formerly known as IMPAC card).

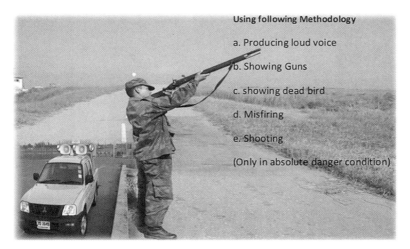

Using following Methodology

a. Producing loud voice

b. Showing Guns

c. showing dead bird

d. Misfiring

e. Shooting

(Only in absolute danger condition)

Figure 6.7 Scaring birds. *Source:* courtesy USDA-APHIS [1].

6.7.1.2   Gas Cannons

Gas cannons (or "exploders" as they are known in the USA) are mechanical devices that produce loud banging noises by igniting either propane or acetylene gas [14].

Gas cannons have the following features:

- they produce 130 dB, which is high enough for bird harassment;
- they are fired at fixed intervals from a control room with the aid of closed-circuit television cameras;
- since birds normally feed and roost in dusk and dawn, propane cannons are fired in these times;
- they must not have a permanent site but should be relocated frequently to avoid bird habituation;
- they are not successful in dispersing all bird species; however, if used together with other depredation techniques, they will be effective for gulls, blackbirds, waterfowl, pheasants, and other game birds;
- prices for gas cannons range from GBP 165 to GBP 475 (for UK products), while worldwide prices range from US$ 450 to US$ 1400;
- costs vary depending on whether it is a single, double, or multi-bang cannon.

Figure 6.8 illustrates a propane cannon used at Baltimore-Washington airport.

6.7.1.3   Bioacoustics

Bioacoustics is simply an artificial reproduction of bird calls or a bird annoyance sound. The technique uses broadcasts of recorded bird distress calls. It may be attached to static or mobile equipment. The early static arrangements on British airfields were very simple. Some fixed loudspeakers, arranged at suitable intervals alongside the runways, broadcast the appropriate calls chosen by the air traffic control (ATC) personnel from the control tower [15].

The main features of bioacoustics techniques are discussed here.

- Static, free-standing systems can be used on smaller areas. However, a louder volume may be needed to cover the area effectively. The calls are broadcast for about

Figure 6.8 Propane gas cannon at Baltimore Washington Airport. *Source:* courtesy USDA-APHIS [3].

90 seconds from a stationary vehicle approximately 100 m from the target flock. Such a system has two disadvantages: constant exposure to a sound originating from a fixed location will encourage habituation and it will cause a noise nuisance to adjacent areas.

- A vehicle equipped with a sound system is used to broadcast distress calls (Figure 6.9).
- The sound system produces noise levels up to 110 dB and a frequency response between 12 000 and 14 000 Hz.

Figure 6.9 Mobile patrol equipped with scaring sound source. *Source:* courtesy IBSC [45].

- The USFWS must be consulted before use if it has an impact upon a federally listed threatened or endangered species.
- The vehicle must be driven close to the birds, with 100–200 m as maximum distance.
- The distress call is to be played for 15–20 seconds, and if the birds do not respond within 20 seconds, the call is played again. If they have not moved by the third attempt, it means that this dispersing method is not successful and other methods have to be used [4].
- Though not all birds are affected by bioacoustics, gulls, starlings, blackbirds, and crows are.
- Recorded distress calls of different bird species will frighten other bird species. As an example, the playback of a peregrine falcon call dispersed gulls at Vancouver International Airport.
- The reaction of birds to such distress calls is to fly toward the source of the call, to circle it, and then move away.
- Bioacoustics are the most effective and cheapest ways of dispersing birds from airfields, once the equipment has been bought and staff trained.

Gloucestershire Airport in the UK has been playing Tina Turner songs from loudspeakers mounted on a van. Apparently, this is successful as the birds are scared away [16].

Moreover, Panama's Tocumen International Airport has experimented with firecrackers and noise guns, but birds have become accustomed to them and no longer respond to the noise.

A low-tech approach to scare birds is to use an unmuffled motorbike (a motorcycle with the muffler/silencer removed). These emit noise from the motor that scares off birds near the runway. This technique is used at NAIA, Philippines. However, birds do get used to this noise. They fly away to the other side of the runway, hover, and then return to their original territory.

6.7.2 Visual Techniques

6.7.2.1 Lasers
A laser is a non-lethal bird repelling technique, which has the following features:

- it is a harmless visual repellent;
- it is effective over long distances (up to 1000 m);
- it is silent and can be safely used in and around structures;
- it offers greater directivity and accuracy over distance;
- it is effective in dispersing birds from bodies of water;
- birds will not get used to its threat, unlike some deterrent devices;
- it can be used to disperse birds roosting on or near runways.

Under low-light conditions, such as sunrise and sunset, and in overcast, rainy, or foggy weather conditions, lasers can be effective in dispersing birds such as geese and cormorants. Effectiveness decreases or even completely vanishes in daylight conditions. Moreover, the use of lasers in an airport environment requires special caution.

Figure 6.10 illustrates both a laser pistol and a laser cannon as well as showing the effective distance for a laser beam. Handheld laser devices are silent and completely

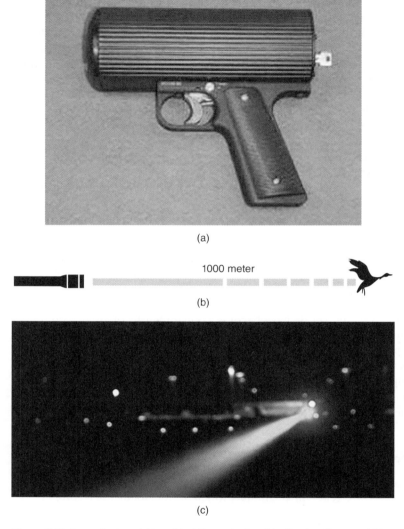

(a)

1000 meter

(b)

(c)

Figure 6.10 Laser devices: (a) hand-held laser device; (b) effective distance; (c) laser cannon. The laser beam can be generated from two sources; namely, handheld sets and cannons. *Source:* courtesy USDA-APHIS [1].

portable devices. They project a 1-inch diameter red or green beam. Birds react to the beam as if they are approaching a car, so they flee the area.

Laser cannons have proved successful in the following test cases under low ambient light conditions [17]:

- Canada geese avoided laser beams for over 80-minute periods, with 96% of the birds moving to untreated areas;
- several thousand double-crested cormorants abandoned roosting at night after three nights of treatment;

- waterfowl species, wading birds, gulls, vultures [47], and American crows avoided the laser beam during field trials; however, its response depended upon context and species.

To a great extent, lasers are successful as avian repellents in several countries, including France, the USA, and the UK [14]. However, further studies are needed to evaluate the response of each species to different parameters like laser power, wavelength, and beam type.

6.7.2.2 Falconry

Falconry is generally defined as fighting birds with birds. Falconry is effective only when used in combination with other frightening techniques (Figure 6.11). Falconry was practiced with good results in some countries, such as Scotland, Canada, and Spain in addition to the USA.

A falconry program incorporates both falcons and several other species of birds of prey. When birds of prey are deployed, they scare and thus disperse birds from the airfield [18].

As an example, at John F. Kennedy International Airport (JFK, New York), falconers daily fly their birds (mostly falcons and hawks) in the airport. Typically, a falcon does not attack and kill target birds but it simulates hunting by chasing a lure swung from a leash by the falconer. In addition to falconry, other methods for dispersing birds, such as pyrotechnics, amplified distress calls, and sometimes a shotgun with live ammunition, may be used simultaneously [19].

However, falconry has the following limitations:

- it is very expensive and labor intensive; since a staff of at least two full-time, well-trained personnel is required to train, operate, and care for the falcons [12];
- several falcons must be available to have at least one bird ready always;
- there is some potential for bird strikes with the falconry birds themselves;
- falcons can be flown only during daylight hours;
- they are limited during high winds, extreme temperatures, rain, and fog;
- most birds of prey used in falconry programs are unsuccessful in dispersing large birds such as waterfowl (particularly geese), as well as other birds of prey.

Figure 6.11 Falconry. *Source:* courtesy IBSC [45].

6.7.2.3 Dogs

Trained dogs like border collies normally keep unwanted geese off airports. Others trained dogs are used to chase and disperse several bird species from the airfields of an airport (Figure 6.12). Border collies respond well to whistle and verbal commands. They act in a strong, predatory manner to scare birds off the runways and taxiways with very limited risk to themselves.

A single border collie together with its handler can keep an area nearly of 50 km^2 free of larger birds. The first commercial airport in the world to employ a border collie was Southwest Florida International Airport in 1999.

Border collies [1, 46] prove effective at dispersing geese, waders, and wildfowl, but ineffective with other species that spend most of their time flying or perching, such as raptors and swallows and especially gulls. Even for geese, a dog can remove them from the airfield but not from water. Generally, all types of dogs will not hurt or even touch the birds since they cannot catch them.

Figure 6.12 Scaring dogs. *Source:* courtesy USDA-APHIS [1].

The disadvantage of using dogs as a predator is its need for thorough training to become highly obedient. The dog must always be under the control of a trained person. Moreover, the initial costs of such dogs are high, covering the purchase of special dogs and their training; other costs are food and veterinary expenses.

6.7.2.4   Scarecrow

Scarecrows are the traditional method used to scare birds. Some of them mimic the appearance of a predator and so cause birds to fly away to avoid potential predation (Figure 6.13). However, many scarecrows are human-shaped effigies, usually constructed from inexpensive materials like black plastic bags and attached to wooden stakes. Such motionless devices either provide only short-term protection or are ineffective.

To maximize the effectiveness of scarecrows, they may be fitted with loose clothing and bright streamers that move and create noise. Also, their location must be changed frequently. Recently, several types of moving, inflatable human effigies have become commercially available worldwide. One of these is the Scary Man®.

One of the most recent design is the flashing hawkeye, which is a ground-mounted wind-powered bird scarer [55]. It is a constantly spinning visual deterrent with large multi-angled mirrors. These give a powerful flash of reflected light from the Sun or even a full Moon.

6.7.2.5   Human Scarer

Human activity can disturb birds from specific areas either intentionally by direct harassment or indirectly, say through leisure activities. The simplest way of frightening

Figure 6.13 Scarecrow. *Source:* Reproduced with permission of scaringbirds.com [55].

birds involves a human standing in full view of the birds and imitating its wing beats by raising and lowering the arms in a frequency of 25 per minute.

A rather costly human scaring method is to employ a four-wheel motorcycle for quick access to the airfield. The costs for a human scarer working a six-day week from approximate dawn until dusk reached GBP 17–33 per person per hectare. Such a technique has been used and led to a reduction in goose numbers in some scaring areas, but it was too expensive.

6.7.2.6 Radio-Controlled Craft

Radio-controlled craft have been used to disperse birds since the early 1980s in both civilian and military airports. The radio-controlled airplane can direct the flying birds to a new flight path away from the airfield. Moreover, for airports located on or close to seas or oceans, radio-controlled boats can harass existing birds.

An example of a radio-controlled aircraft is the Robo-Falcon™ (Figure 6.14a), which may be used – after special training for the personnel involved – in conjunction with other depredation techniques for dispersing birds from airports [1]. For airport's office and housing areas, radio-controlled vehicles can be used effectively to disperse birds from grassy areas since the use of noisy methods such as pyrotechnics or gas cannons in such areas is not desirable [4].

At Whiteman Air Force Base, Missouri, balsa wood radio-controlled aircraft are used to keep the airfield clear of raptors and other large birds. They also proved effective at dispersing the base's redwing blackbird roost.

The Korean Atomic Energy Group and LIG Nex1 developed the world's first bird strike defense robot. It is a six-wheeled unmanned ground vehicle (UGV) that uses both acoustics and a laser to scare birds (Figure 6.14b).

6.7.2.7 All-Terrain Vehicles (ATV)

ATVs are used for dispersing birds in the aircraft operating area. They may be used in conjunction with other pyrotechnic methods. Though they are four-wheel drive vehicles, they are sometimes difficult to operate in some terrains [4].

6.7.2.8 Pulsating Lights

The Pulselite system is an effective system that substantially reduces bird strikes and significantly increases the visibility of aircraft (Figure 6.15). It was tested aboard the fleet of Alaska Airlines, Horizon Airlines, and Qantas Airways for 10 years. It proved that bird strikes had been reduced by 30–66%. These reductions result in increased profitability, enhanced safety, improved customer service, and higher customer satisfaction ratings [20].

6.7.2.9 Scaring Aircraft

Frightening paints may be added to either of conventional aircraft or helicopter that scare birds [54]. Flying of such frighteningly painted aircraft at Vancouver Airport successfully deterred gulls, ducks, and geese [21].

6.7.2.10 The Robotic Peregrine, Hawk and Falcon (Robop and Robird)

Robotic birds are used to repel birds from airfields. Robop (Figures 6.16 and 6.17) is designed by Robop Limited (located east of Edinburgh, UK), and successfully installed

(a)

(b)

Figure 6.14 (a) Radio-controlled aircraft. (b) Radio-controlled unmanned ground vehicle. *Source:* courtesy USDA-APHIS [1].

in 16 countries so far. It is designed to have the shape, behavior, and sound of a hunting peregrine falcon. It thus scares birds such as seagulls and pigeons to search for new territory. Robop can move its head and wings. It also emits four peregrine calls to deter birds on airport runways. It has two advantages: it causes no harm to birds and it can withstand harsh weather conditions for many years with minimal maintenance.

Alternatively, remote-controlled falcons and eagles (sometimes called "Robirds") are designed to fly over airfield areas, scaring off intruders [22]. The Robirds project is being developed by Clear Flight Solutions in both the Netherlands and Canada.

In the Netherlands, there are two Robirds for scaring small and big birds. The smaller is a falcon while the larger is an eagle (Figure 6.18). The falcon robot has a body length

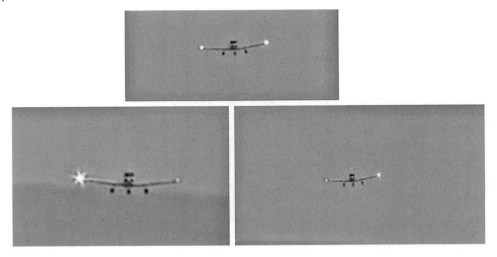

Figure 6.15 Pulselite system.

Figure 6.16 Stationary Robop Falcon. *Source:* courtesy Robop Limited [53].

of 23 inches and a wingspan of 47 inches while its flight speed is around 50 mph. The eagle robot is 46 inches long and has an 86 inch wingspan. Both are remote-controlled for now. Manufacturers hope to make them autonomous in the future. Both birds are 3D printed using a glass fiber and nylon composite material and painted to make them look as realistic as possible.

Figure 6.17 Robop Falcon in Scipol Airport. *Source:* courtesy Robop Limited [53].

Figure 6.18 Flying Robirds. *Source:* courtesy USDA-APHIS [1].

In Canada, the Robirds are robotic birds of prey that employ a flapping wing motion during flight, the same as real birds.

6.7.2.11 Corpses

Deploying replicas or actual dead birds conserved with formaldehyde (Figure 6.19) will help in repelling birds, especially gulls. When birds approach the corpse, they see the

Figure 6.19 Dead birds. *Source:* courtesy USDA-APHIS [50].

unnatural position of the bird. They feel danger and leave the area. This approach has been frequently used in attempts to deter gulls from airports [21].

6.7.3 Lethal Techniques

Some bird species may grow accustomed to bioacoustics and pyrotechnics techniques. Moreover, other species are so dangerous that they influence the safety of flight vehicles and personnel. In such cases, lethal means must be adopted, after getting a federal depredation permit, especially for any protected birds. Generally, legislative protection of wildlife prohibits the intentional killing, injuring, or damaging or destroying of the nest or eggs of most bird species. However, some birds like European starlings, house sparrows, and rock doves/domestic pigeons are not federally protected in the USA and so do not require any federal depredation permit. While, bald, and golden eagles, or their nests and eggs, are federally protected [4]. In the UK the Wildlife and Countryside Act issue licenses for using lethal techniques for bird control management.

6.7.3.1 Shooting

Firearms are greatly restricted and are used only after all other wildlife control methods have failed to repel the birds. Firearms are also used in cases where the immediate removal of problematic birds is necessary. Both the FAA and USDA-APHIS support live-ammunition shooting as an "effective practice" for bird population reduction.

Figure 6.20 Bird shooters using pistol and gun. *Source:* courtesy IBSC [45].

Airports around the world have attempted to decrease bird strikes using shooting. Most airports worldwide, including JFK, have adopted shooting birds and oiling eggs to decrease the bird population.

Figure 6.20 illustrates typical shooters using a pistol and a gun for shooting birds. Along the southern airport boundaries of JFK, where gulls often cross the airport, two to five shooters were stationed. Shooters, wearing blaze-orange vests, direct their guns away from the airport but towards flying gulls that came within range (about 40 m). All shooters have permissions from the USDA-APHIS or PANYNJ. Table 6.3 illustrates data for a case study looking at JFK, where a program was conducted to reduce gull strikes. The program focused on strikes in the period 1991–2002, mainly by laughing gulls as well as three other gull species [36]. The critical period in the year was from May to August since laughing gulls are extensively present due to the nesting colony in Jamaica Bay next to the airport. The shooting was conducted for 31–62 days by two to five people. Table 6.3 gives the annual numbers for aircraft movements, shooting days, shots fired, and number of laughing and non-laughing gulls killed [23].

Table 6.3 Shooting data in a case study of JFK International Airport [23].

Year	Aircraft movements	No. of days	Shots fired	Laughing gulls killed	Non-laughing gulls killed	Total gulls killed
1992	323 448	61	31 183	11 847	1 619	13 466
1994	343 599	31	12 510	3 688	293	3 981
1996	355 214	34	7 651	1 970	293	2 263
1998	343 429	43	9 008	2 920	298	3 218
2000	345 311	61	15 010	3 606	613	4 219
2001	292 367	60	13 753	3 194	629	3 823

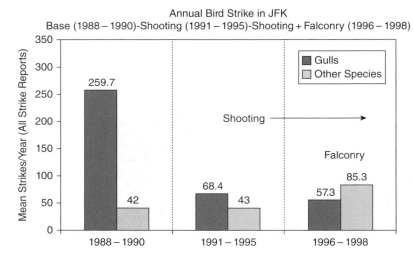

Figure 6.21 Comparison of shooting and falconry with the base number of bird strikes at JFK International Airport. *Source:* courtesy IBSC [23].

In cases of emergencies, the USFWS allows airport authorities to kill, capture, and relocate up to 10 migratory bird species. "Emergency" here means either an immediate danger to public safety or an immediate hazard to aircraft.

The effectiveness of a combination of shooting and falconry is illustrated in Figure 6.21. This figure shows the number of aircraft striking gulls only and other species at JFK in the period 1988–1998.

- During 1988–1990 there was neither shooting nor falconry.
- During 1991–1995, with shooting only (no falconry), there were drastic reductions of strikes with gulls. The statistics for other birds were nearly the same as 1988–1990.
- For the period 1996–1998, when both shooting and falconry were used [24], the numbers of strikes for gulls continued to decrease. However, the number of strikes from other species increased to the double the comparable numbers in the period 1988–1990 (where neither falconry nor shooting was employed) and double the numbers in the period 1991–1995 (when shooting only was done).

6.7.3.2   Live Trapping
Live trapping is mostly used against state and federally protected and high-profile bird species, and the birds are then relocated away from the airport. Live trapping is time-consuming and relatively costly. Live trapping may use mist nets, cage traps, cannon nets, or large funnel-shaped lead-in traps. Live traps may be used to capture sedentary birds such as pigeons and house sparrows (Figure 6.22).

6.7.3.3   Removal of Nests and Young
Hazardous bird species such as Canada geese, mute swans, gulls, and others must not be allowed to nest on airport property. Any nests with eggs found at an airport can be destroyed by breaking the eggs and removing nest materials. However, local, regional, and national wildlife laws must be checked before removing, altering, or interfering

Figure 6.22 Pigeon trap. (Source JFK) [25]. National Academies Press.

with any bird nests in any way. If the nest has been abandoned or no eggs have yet been laid, it can be removed or destroyed as needed.

Under the US Migratory Bird Treaty Act (MBTA) one is not allowed to kill any migratory birds or destroy nests that have eggs or brooding adults in them without permission from the USFWS. If permission is granted, the removal of nests, eggs, chicks, and young must be done every two weeks. However, this technique is time-consuming. However, nests of invasive birds like European starlings or house sparrows are not protected.

A license is required in the UK. Laws in other countries may vary. The bird species must be defined and local laws must be consulted first.

Water spray has been used to prevent birds from roosting or nesting both in urban and agricultural areas. Water cannons and sprinkler systems can be used to control "pest" birds (Figure 6.23). Water penetrates bird feathers, resulting in their death as feathers become wet and their body temperatures drops. It is also successful when applied at night in roost locations. Airport fire trucks can be used to damage swallow nests before egg laying.

As an alternative method, nests on the ground may be removed by hand while those in trees can be dislodged with a telescopic pole. If the nesting material is scattered, it will discourage rebuilding.

Finally, an effective way of repelling birds is to apply a specific gel which will not allow them to perch or nest. This is waterproof, and almost odorless and colorless. It is also effective in high temperatures.

6.7.3.4 Egg Manipulation

Egg piercing, shaking, or oiling can be very effective in controlling birds by preventing hatching if lethal control is not an option (Figure 6.24). The pricking of eggs with a needle allows bacteria to enter the egg and desiccate its contents. However, some

Figure 6.23 Water cannon used to remove cliff swallow nests under support structures. (*Source:* SLC) [25]. National Academies Press.

Figure 6.24 Egg oiling. (*Source:* SLC) [25]. National Academies Press.

pricked eggs may still hatch. Egg oiling is a cheaper and more effective method of egg control. It involves coating the egg shells with oil such as liquid paraffin. JFK adopted oiling eggs to decrease bird population.

6.7.4 Chemical Repellents

Chemical repellents, toxicants, and capturing agents are listed in Hygnstrom et al. [27]. They include polybutene, anthraquinone, methyl anthranilate, aminopyridine, and aluminum ammonium sulfate.

In the USA, all chemical repellents must be registered with either of the Food and Drug Administration (FDA) or the US Environmental Protection Agency (USEPA) before they can be used to manage wildlife at airports. Moreover, they must also be registered in each state. In several other countries, including the Netherlands, chemical repellents are neither used nor even experimented due to toxic features [26].

Chemical products can be applied to perching structures, vegetation, turf grass, and standing waters in the airport. Moreover, they are sometimes applied to temporary pools of standing water on airports to repel birds until the water evaporates (Figure 6.25). However, a long-term solution for such a problem is preferred. The drainage system should be improved to avoid standing water after significant rain events.

The following sections cover the chemical repellents, listed by active ingredient, that are presently available for use on airports. The overall conclusion, however, is that the use of chemicals must be combined with other techniques to chase birds away.

6.7.4.1 Polybutene

Polybutene may be found in either liquid or paste formulations. If applied to perch structures it makes the bird feel uncomfortable due to its sticky nature. The birds will move to an untreated surface. Under normal conditions, the effective lifetime of polybutene is 6–12 months but reduces substantially in dusty environments.

Polybutene is effective for controlling pigeons and preventing raptors from perching on antennas. However, for smaller birds such as sparrows, it is less effective since they require only a small area for perching and may find areas with no repellent.

6.7.4.2 Anthraquinone

Anthraquinone is usually available in liquid formulations and applied by sprayer to the vegetation [27, 28] (Figure 6.25). It is registered as a chemical used for repelling geese from turf since it acts as an aversion repellent to birds. If they ingest food treated with anthraquinone, birds will become slightly ill and develop a post-ingestion aversion to such a treated food. Anthraquinone is used only in high-risk areas of runways as well as bird grazing areas. It is not registered in the UK.

6.7.4.3 Methyl Anthranilate

Methyl anthranilate is tested and registered by Homestead Air Reserve Station in Florida as a feeding repellent for birds (geese, gulls, waterfowl, and starlings) on turf and

Figure 6.25 Chemical repellents applied to temporary pools of standing water. *Source:* courtesy USDA-APHIS [1].

golf courses [1]. It is not registered in the UK. It is a non-toxic active compound in ReJeX-iT. Birds have a taste aversion to methyl anthranilate.

Similarly to anthraquinone, methyl anthranilate products are liquid formulations that are applied by sprayer to the vegetation. Its effectiveness in repelling geese varies depending on growing conditions, rainfall, mowing, and availability of alternate feeding areas.

6.7.4.4 Naphthalene
Naphthalene is a repellent that works on the sense of smell. It was tested at airfields in the UK. It was applied to the field as "mothballs." However, the results were contradictory [12].

6.7.4.5 Avitrol
Avitrol is an example of a toxic repellent. It is registered for repelling several bird species, including pigeons, house sparrows, blackbirds, and gulls, from feeding, loafing, roosting, and nesting. When birds eat baits (normally grain) that are strongly treated with Avitrol, then they will have distress behavior. Consequently, it frightens other nearby birds. If a limited amount of bait is used, then birds will be chased away with minimum mortality [4].

6.7.5 Exclusion

Exclusion is the use of physical barriers to stop birds from gaining access to their necessities of life, food, water, or shelter, at or near the airport [29].

6.7.5.1 Netting
Netting is used to prevent birds from gaining access to food or roosting areas. It is generally available in rolls having a width of 3–4 m, lengths of 16–1200 m, and a mesh

opening of 3–3.5 cm. Netting is used at airport buildings and hangers to cover holes and openings as well as to prevent birds from nesting or perching in the beams and girders of the ceiling support structure.

Netting needs no permission.

6.7.5.2   Porcupine Wire (Nixalite)

Porcupine wire, having the trade name of Nixalite, prevents birds from perching and roosting on flat surfaces. It is stainless steel strips that feature needle-like wire prongs, having 0.5 cm base and either 5 or 9.5 cm length. It has proved its effectiveness for all bird species and on all types of surfaces. However, it is expensive and unless installed on all ledges birds may move to a new location nearby. Moreover, some birds having sufficiently long legs, such as red-tailed hawks, can perch on Nixalite-covered surfaces.

No permission is needed for its usage.

6.7.5.3   Bird-B-Gone

Bird-B-Gone is strips of durable plastic porcupine wire that is less expensive than Nixalite. It must also be installed to all ledges. Some birds are not deterred.

Permits are required.

6.7.5.4   Avi-Away

The Avi-Away system consists of a control unit and an electric cable installed along the area to be protected. When a bird lands on the cable, it completes an electrical circuit and receives a mild shock and a recorded bird alarm call, which disperses other nearby birds. Avi-Away can permanently exclude all birds from installed locations.

This system needs a permit and periodic maintenance.

6.7.5.5   Fine Wires (Large-Area Applications)

A grid of fine stainless-steel wires having diameters less than 0.5 mm is installed some 1 m above sources of food and water (ponds, standing water, and landfills at airports).

Wire grids may be installed over flat roofs to deter gulls. Strands of wire should be installed in parallel and on a horizontal plane. Wires spaced at 6-m widths reduce the numbers of ring-billed gulls, while wires spaced at 12 m are effective at deterring herring gulls.

Wires must be checked regularly. No permission is required.

6.7.5.6   Bird Balls™

Bird Balls are hollow-plastic balls which prevent birds from landing on standing water. They are alternatives to netting and wires. When installed, they cover the entire water surface, and thus all birds see from the air is what appears to be a solid surface and they fly elsewhere searching for water.

Advantages are an easy installation, they do not break or tear, they are effective in all weather conditions, and need little maintenance. On the other hand, they are more expensive than netting and wires and effective on standing water only.

No permission is required.

6.8   Habitat Modification or Passive Management Techniques

The essence of habitat modification is to modify the ecological character of the airfield by eliminating sources of water, food, and shelter so that the area ceases to be an attractive place to birds [30].

Habitat management is both effective and environmentally friendly. It involves two processes:

- identifying the attractive features;
- implementing changes that either removes the attraction (food, water, and shelter) or deny access to it.

Though the passive management is costly, its solutions are long term, and, once competently applied, they can reduce bird strikes for many years. However, full application of passive methods will need smooth cooperation with owners of the surrounding areas, local administration, and business interest groups.

6.8.1   Food Control

Birds are attracted by the presence of fish, frogs, insects, rodents, edible waste, fruit-producing trees, seed-producing vegetation, green weeds, grass, aquatic vegetation, agricultural grains, insects, and earthworms.

The objectives of food control are:

- to prohibit bird feeding;
- to eliminate all human food waste;
- to manage any grass areas within the airports' boundaries;
- to manage all types of agricultural activity, including shrubs and trees, on or near an airport;
- to remove all seed-bearing plants;
- to use insecticides to eliminate food sources for insect-eating birds.

Details for food control methods are given now.

Waste management (refuse collection, landfill sites, and garbage dumps) may be achieved by making the site as unattractive as possible, using one or more of the following methods:

- close the existing dumps that attract birds;
- prohibit or restrict the establishment of new sites that accept organic waste close to airports (new dump sites be should not be closer than a 13 km circle centered on the ARP);
- cover waste material with rejected compost, auto-fluff, and tarps;
- fencing;
- netting (Figure 6.26);
- utilize the cell method rather than the area method; where the cell method exposes only a small portion of the working face at any one time, thereby minimizing the exposed wastes for birds;
- incinerate waste;

Figure 6.26 Netting over landfill close to an airport. *Source:* courtesy IBSC [13].

- operate the landfill either as a pit or trench;
- dump waste at night or during non-flying periods;
- use overhead wire barriers (to discourage gulls in particular);
- adopt fully enclosed waste-transfer facilities and sites which handle inorganic refuse such as construction and demolition;
- relocate landfills that do not meet FAA guidelines criteria.

If relocation is not feasible, adopt other environmentally sustainable and non-lethal methods. Lethal reinforcement may be only used as the method of last resort.

Grass management: fields of short grass are used by hazardous species such as gulls, waterfowl, starlings, and lapwings for feeding. Long and dense grass has proved effective in reducing the risk of bird strike since it:

- provide difficulties for birds to find food such as worms and insects;
- reduces the visual ability of birds to detect predators in the surroundings;
- reduces the necessary space for birds to achieve the wing beat needed for takeoff.

Long grass programs are being increasingly adopted in the USA. The height of airfield grass must be in the range of 17–35 cm, as per USAF instructions. Even lower heights (15–25 cm) are instructed by the USDA-APHIS for JFK. However, long grass attracts rodents, and under some circumstances may lead to an increase in raptor populations.

When it is unclear which food sources (long or short grass) are the major attractants, it is also difficult to determine how to eliminate them [29].

Herbicides and growth retardants: herbicides are applied to keep broad-leafed weeds to a minimum on the airfield as they attract various birds. Generally, growth retardants should be tested on small test pilot projects before use in large areas [4].

Planting bare areas: since birds use bare areas for resting, reducing those close to airfields will reduce bird strike possibilities. Moreover, birds pick up grit to crush seeds to help their digestion. Bare areas are planted with grass as linear plantings such as hedgerows and tree lines should be avoided. Removal of needless groups of trees or hedgerows is recommended.

Fertilizing: fertilizers are used to stimulate grasses to the heights suggested by authorities.

6.8.2 Water Control

The objectives of water control are:

- to eliminate or modify habitat features, including open areas of water, as even transient water accumulation on uneven pavements that attract birds must be managed;
- to manage landfill and other waste disposal sites that attract large numbers of birds.

Details for water control methods are given now.

Airports may include wetlands, ditches, ponds, creeks, rivers, and lakes. Such water sources attract waterfowl and shorebirds as well as other species. In addition to the above permanent water sources, there are areas where water collects for short periods of time (say after rainstorms or during spring snow melt). Such standing water at airports should be removed or modified immediately.

Water-habitat management requires both long-term permanent solutions as well as short-term ones.

Long-term solutions include:

- realignment or replacement of ineffective ditching systems with buried drain pipes;
- artificial and natural ponds should be eliminated through infilling, grading, and improving drainage;
- relocation of stormwater ponds to safe areas.

Short-term solutions include:

- creating unappealing water habitats, perhaps by cleaning and removal of aquatic vegetation, as well as making their banks deep and steep;
- altering habitats to exclude specific problem species by regularly cutting vegetation.

Here are four proven methods that prevent birds from either landing or swimming on water surfaces:

1) overhead systems using wires made of metal, nylon, or monofilament – a grid of 2.5–6 m will stop most gulls while a grid of 3–4 m system will stop most waterfowl;
2) fine netting (Figure 6.27);
3) flagging tape;
4) plastic balls, which float on water, make the water inaccessible to birds (Figure 6.28).

6.8.3 Shelter Control

Birds need shelter for breeding, resting, roosting at night, and protection against predators. Birds will find natural or human-made shelter in the following areas:

Figure 6.27 Netted drainage. *Source:* courtesy IBSC [13].

Figure 6.28 Floating plastic balls to prevent waterfowl and other birds from accessing the water. *Source:* courtesy IBSC [13].

- shelter habitats such as towers, old hangers and forested areas on the airfield;
- shrubs, open short-grass fields, and trees;
- crevices and holes;
- vents and ducts;
- roof ledges of buildings;
- signs, airport lighting fixtures, and buildings

- water bodies – drainage ditches and sewage lagoons;
- overhead wires.

Some details for shelter control methods to reduce bird attractants are given here.

6.8.3.1   Managing Reforested Areas

It is important to select suitable locations for artificial forest areas that will not add to bird strike problems close to airfields. It is also recommended to select appropriate trees for such forests that will not become a perfect place for roosting. As an example, pine trees are not recommended. Moreover, for the existing forests, it may be necessary to remove some or all of their trees or replace them with airfield turf to discourage birds' roosting behavior.

Should a stand of trees contain birds protected by the MBTA, the Bald and Golden Eagle Protection Act, or any other protected species, the USFWS should be consulted to determine whether a permit is needed before tree removal [4].

6.8.3.2   Landscaping

It is not recommended to plant trees or shrubs in the airfield areas since it will increase bird populations and movements [31]. The following recommendations must be followed when planting trees/shrubs:

- select trees that do not produce fruit which provides food (such as berries) for birds;
- avoid trees with dense foliage since they provide the bird with all their needs (food, shelter, and nesting areas).

Birds can usually be stimulated to move by pruning and thinning trees and shrubs to open the canopy. In some situations, it may be necessary to remove all the plant.

In summary, Table 6.4 summarizes the different methods used in the active and passive control of bird species [20]. Effectiveness is variable, and a combination of methods is often necessary.

Table 6.5 demonstrates the effectiveness of the most important methods for bird management at airports. Small flocking passerines include starlings, swallows, and some fringillid finches.

6.9   Air Traffic Service Providers

ATS assures aircraft safety both on the ground and during flight. It provides instructions for controllers of the following activities:

- arrival and departure;
- en-route flight.

ATS is are responsible for risk management activities including wildlife hazard and strike minimization [23].

6.9.1   Controllers and Flight-Service Specialists

Controllers and FSS have numerous responsibilities regarding pilots and airport personnel for bird strike prevention. Details are listed below.

Table 6.4 Active and passive control methods most often used for various bird species [20].

Bird species	Appropriate technique
Black-crowned night heron	complete enclosure (siege) perimeter fencing automatic exploder pyrotechnic distress calls lights
Gull	complete enclosure (siege) perimeter fencing automatic exploder pyrotechnic distress calls water spray
Great blue heron	complete enclosure (siege) overhead lines (wires) perimeter fencing automatic exploder pyrotechnics lights
Green-backed heron	complete enclosure (siege) perimeter fencing automatic exploder pyrotechnics
Snowy egret	complete enclosure (siege) overhead lines (wires) perimeter fencing automatic exploder pyrotechnics
Blackbird	complete enclosure (siege) perimeter fencing automatic exploder pyrotechnics distress calls
Merganser and diving ducks	complete enclosure (siege) perimeter fencing automatic exploder
Belted Kingfisher	complete enclosure (siege) automatic exploder pyrotechnic
Dipper	complete enclosure (siege)

(Continued)

Table 6.4 (Continued)

Bird species	Appropriate technique
Cormorant	complete enclosure (siege)
	automatic exploder
	pyrotechnics
Tern	complete enclosure (siege)
	overhead lines (wires)
	pyrotechnics
Osprey	complete enclosure (siege)
	overhead lines (wires)
Pelican	complete enclosure (siege)
	perimeter fencing
	automatic exploder
	pyrotechnics
Grebe	complete enclosure (siege)

Duties concerning pilots:

- provide pilots with any information about possible bird activities;
- update pilots with recent information concerning bird activity at or near airports;
- transfer any information concerning birds from Automatic Terminal Information Service (ATIS) and Notice to Airmen (NOTAM) to pilots;
- encourage pilots to report all bird strike accidents/incidents after bird strikes or near misses;
- in the event of bird strike threat, provide different options to pilots which include:

  - takeoff delay
  - use of different runways for landing and takeoff;
  - employ alternate flight profiles;
  - use alternate routes and altitudes;
  - reduce aircraft operating speed.

Duties concerning airport personnel:

- controllers are also instructed to continue transmitting warnings as long as the hazard is present [23];
- inform appropriate airport personnel about bird activity at airports;
- advise shift replacements about current bird activity over and near the airport;
- ensure that active bird control methods pose no threat to aircraft operations.

6.9.2 Terminal Controllers

Terminal controllers responsible for the arrival and departure of aircraft play a vital role in protecting aircraft from bird strikes [32]. They transfer information regarding bird

Table 6.5 The effectiveness of the most important methods for bird management.

Method		Effectiveness for small flocking passerines	Effectiveness for corvid birds	Effectiveness for gulls
Active management	Mobile patrol	T/–	T	T
	Acoustic repellents			
	Propane cannons	T	T	T
	Playbacks of distress or alarm calls	T/–	T	T
	Visual repellents			
	Flares	–	T	T
	Radio-controlled models	?	T	T
	Chemical repellents	–	T	T
	Falconry	T	T	T
	Trained dogs	–	T	T
	Physical elimination of birds	T	T	T
	Capture and removal of birds	–	T	T
Passive management	*Food*			
	Grass	+	–	+
	Bird-proof waste containers	–	+	–
	Dumping grounds	–	+	+
	Fruiting trees and shrubs	+	+	–
	Crop fields	+	+	+
	Earthworms	+	–	+
	Insects and other invertebrates	+	–	–
	Water	+	+	+
	Shelters	+	+	–

The following symbols are used to identify the effectiveness of the method: (+) effective, (–) ineffective, (T) are temporarily effective.

activities either directly to flight crews of arriving and departing aircraft or indirectly, through tower controllers, for pilots of aircraft still on the ground, and NOTAMs, for pilots of other aircraft who plan to operate at or near an airport. These communications increase the awareness of flight crews and reduces the possible severity of bird strikes.

6.9.3 Tower and Ground Controllers

Tower controllers use both sighting and binoculars to identify the full details of any flock of birds; namely, their type, size, location, and flight direction. They also confirm other sightings by airside staff and verbal pilot reports [32, chapter 9].

6.9.4 Flight Service Specialists (FSS)

FSS provide pilots and decision makers with vital and timely information regarding bird management.
FSS duties are:

- to detect and identify details of any approaching flock of birds (type, numbers, location, and flight direction);
- to brief airport managers of necessary work due to potentially hazardous birds;
- to convey information among pilots, airside workers, and bird-management personnel;
- to suggest alternative routings and, when appropriate, recommend that flights not be attempted.

6.9.5 Pilots

Most pilots will pursue their career without encountering a significant bird strike event. The term "significant" implies severe damage and risk to the aircraft. Many recommendations are given to pilots for bird strike avoidance and risk mitigation by airliners manufacturers and associations such as Boeing [6] and Airbus [33] as well as TC [29] and the Airline Pilot Association [24].
The general rules for pilots may be stated first [6]:

- rely upon different outboard and onboard sources for detecting flight conditions such as weather. time of day, runways conditions, and any threat of flocking birds;
- avoid or minimize maneuvering at low altitude (either during climb or descent flight phases);
- modify the flight route in the presence of known or anticipated bird activity.

More detailed information will now be provided.

6.9.5.1 Preflight Preparation
1) Observe any bird activity close to the aircraft [32, Chapter 10].
2) During the preflight walk-around, check for any sign of birds nesting in any cavities of the airframe and inlet/outlet of engines. Bird droppings and straw on the ground close to the aircraft prove the existence of nests.
3) Note any bird activities, after receiving the airport information from ATS providers, ATIS, or UNICOM.
4) Heat the windshields if possible as this will increase both its pliability and ability to withstand bird impacts.
5) Before engine startup, review all aircraft emergency procedures if bird strike is encountered, such as rejected takeoff and engine-failure procedures.

6.9.5.2   Taxiing for Takeoff

1) During taxiing for takeoff, the pilot must observe any bird strike activity if informed by ATS providers or other operators [32, Chapter 10].
2) Report any bird activity on both ramps, taxiways, and runways to ATS providers, UNICOM, and other aircraft.
3) Check the length of the active runway and be sure that there are no birds, especially when operating at airports that either do not have ATS providers at all or have limited hours of ATS operation since these airports have no formal bird control/management.

6.9.5.3   Takeoff

The pilot must carefully consider the following points:

1) Check the runway for the presence of birds again before starting takeoff [32, Chapter 10].
2) If an aircraft is taking off in front of him, it may frighten birds which may then fly toward his flight path.
3) Use landing lights (particularly pulsed landing lights) which push birds to fly away from the aircraft.
4) Be sure that aircraft weather radar is not an effective way for warning birds since its power emissions and frequencies are low.
5) If there is bird activity on the runway, use another runway not affected by the birds, wait for wildlife management personnel to clear the birds, or delay takeoff if not sure of flight safety.
6) If a bird strike occurs during the takeoff roll, the decision to continue or abort takeoff should be based on the aircraft's flight manual aborted criteria.
7) If a bird strike occurs, then the aircraft should be inspected by maintenance engineers.

6.9.5.4   Climb

The pilot must follow the following during the climb.

1) Be prepared to adjust climb route to avoid birds [32, Chapter 10].
2) Reduce airspeed to minimize the impact force and possible aircraft damage in case of bird activity.
3) Climb at the maximum possible rate of climb identified by the manufacturer through the altitude bands where birds are flying.
4) Keep speed below 250 knots as well as continuing to use landing lights until exceeding 10 000 ft (AGL), since most bird strikes occur below this altitude.

6.9.5.5   En Route

1) Get information concerning bird activity from ATS providers, ATIS, UNICOM, and surrounding aircraft via appropriate en route radio.
2) Report all bird movements to ATS providers and close aircraft.

6.9.5.6   Approach and Landing

The pilot must follow these instructions.

1) Obtain the latest information regarding bird activities from different sources including ATIS, UNICOM as well as close aircraft.

2) Avoid descent and approach routes over areas that attract birds [32, Chapter 10].
3) Watch any bird activity throughout approach and landing phases and request airport personnel to clear runways from birds (if any), especially in cases of airports which either have no ATS providers or have limited hours of ATS operation.
4) Reduce airspeed to minimize the impact force and possible aircraft damage in areas with high bird activity.
5) Descend at the highest permissible rate (without increasing speed) if birds are reported at certain altitudes, so as to minimize the time of aircraft exposure to possible bird strikes.
6) Delay landing in case of bird strike hazard until conditions become appropriate.
7) Consider a go-around instead of continuing descent and plan a second approach if a flock of birds is seen ahead.
8) Continue the approach and land if birds hit the aircraft.
9) Switch to idle power if descending over water courses, to avoid extended low altitude level flight.
10) Use landing lights during approach and landing to push any birds away from aircraft.
11) Report any bird activity encountered to ATS providers, UNICOM, and close aircraft.

6.9.5.7 Post-Flight

In case of real or suspected bird strike, the following actions must be followed.

- Report the incident to the transport authorities, with all details, photographs of bird remains.
- Collect any bird remains to be sent to the laboratory.
- Maintenance engineers must thoroughly inspect the aircraft prior to its next flight.

6.9.6 Air Operators

6.9.6.1 Introduction

The main role of air operators is reducing the probability and severity of bird strikes through the following three activities:

- applying Standard Operating Procedures (SOPs);
- training of employees in awareness;
- reporting bird strikes.

Bird hazard-specific SOPs should be developed, included, and amended in airline publications, including the following departments/personnel:

- flight operations (pilots/flight crew);
- flight dispatch (flight planning and flight following);
- ramp operations (aircraft ground handling);
- aircraft maintenance sections (airframe, engines, electrical, communication, etc.).

Both flight operations and flight dispatch personnel use common SOPs. Some recommendations listed in airline manuals that help in reducing bird strike probabilities and severities are provided here [32].

6.9.6.2   Air Operator General Flight Planning and Operating Principles

Plan and carry out all flights based on gained experience in the reduction of bird strike risk.

1) Flight plans must be at the highest possible altitudes (greater than 3000 ft AGL) since most bird strikes occur below this altitude.
2) Maximize rate of climb on departure to get above 3000 ft AGL as quickly as possible.
3) Arrange climb to be as close as possible to the boundaries of the airport as bird activity is managed within the airport boundary.
4) Reduce speed at climb and descent to limit the severity of bird strikes.
5) Use landing lights always when flying below 10 000 ft AGL.
6) Avoid planning flight routes over areas that attract birds.
7) Regarding daily plans, the following bird activities have to be considered:

   - all birds fly during the day;
   - some species (such as owls and migratory waterfowl) fly at night;
   - few species fly at dawn and dusk.

8) Concerning yearly plans in Canada, be aware that the critical times for bird strike are March to April, July to August, and September to October.
9) In North America, a significant percentage of the Canada goose population remains in areas close to many airports throughout the whole year.
10) On hot days bird species like raptors and gulls soar to a high altitude.
11) Recent aircraft are so quiet that birds do not hear them and so collide with them. In such cases, the pilot should climb above bird flocks and maintain a safe speed.
12) If bird strike is encountered, then maintenance personnel must inspect the aircraft before the next flight, based on the airline maintenance manual.

6.9.6.3   Flight Planning

During the flight-planning process, all the available information regarding the flight route and possible bird hazards must be reviewed. The review includes the departure airport (for taxiing for takeoff, takeoff, and climb), the flight route, the approach, and the landing airport, as well as the alternative airport in cases of emergency as well as post-flight checks [32].

6.9.6.4   Managing Agricultural Programs in Airfields

The following points must be noted when planting crops in airfields:

1) grain crops should be 1000 ft away from the runways;
2) plant crops that are least attractive to birds, like hay, cotton, and flax;
3) crops planted in airfields should be identical to those planted in the surrounding area.

6.10   Aircraft Design

Aviation authorities, as well as airframe and engine manufacturers, are aware of the risks that bird strikes pose to aircraft and their engines. US Federal Aviation Regulations (FARs) assures the safety of commercial aircraft against bird strikes. Regulations require proof tests of both aircraft and their engines [32], as will be described in Chapter 7.

Different modules of the aircraft or engine (such as wing/tail leading edge, windshield, and fan blades) must not fail if impacted by a designated bird mass and velocity.

Current research and development by aircraft and engine manufacturers focuses on the following three areas:

- certification standards;
- design and material selections of modules;
- impact tests.

6.10.1 Certification Standards

A periodic review for airframe [34] and engine airworthiness standards is performed to enact more stringent impact-strength requirements for both airframes and engines [35]. A summary of these standards will be given here, while the full details will be given in Chapter 7.

6.10.1.1 Airframe Certification Standards

Large Aircraft: numerous requirements are specified for different modules as follows:

- Airframe – designers have to stick to the standards (14 CFR Part 25-571 and EASA CS-25.631) for continuous flight after the impact of a single bird with an aircraft.
- Windshield – it must withstand the impact with a 1.8-kg bird at cruise speed at mean sea level without penetration.
- Pitot tubes – it must be far enough to avoid damage from a single bird impact.
- Empennage – it is necessary to comply with the EASA CS-25 and 14 CFR Part 25 as well as the FAA additional requirement under 14 CFR Part 25-631 for a continued safe flight and normal landing after the impact by a 3.6-kg bird at cruise speed at mean sea level.

Small aircraft: it must maintain windshield integrity if impacted by a 0.91-kg bird based on the standards EASA CS-23.775 and 14 CFR Part 23.775.

6.10.1.2 Engine Certification Standards

Standards FAA 14 CFR Part 33-77 and EASA Airworthiness Code CS-E 800 "Bird Strike and Ingestion" for both multiple and single bird engine ingestions into a single fixed-wing aircraft engine state that at a typical initial climb speed and takeoff thrust:

- if a single bird of weight 1.85–2.5 kg (based on engine inlet area) is ingested into the engine, it will neither fail nor catch fire; the engine should also sustain 50% of the developed thrust for at least 14 minutes after ingestion without moving the thrust lever for 15 seconds post impact;
- following ingestion of a single bird of maximum weight 1.35 kg, thrust or power should not decrease by more than 25%; the engine must keep running for at least 5 minutes without reaching a hazardous condition;
- simultaneous ingestion of up to seven medium-sized birds having weights of 0.35–1.15 kg, depending on the engine inlet area, shall not cause the engine to fail suddenly and completely; the engine should continue to deliver thrust for at least 20 minutes after ingestion;

- simultaneous ingestion of up to 16 small-sized birds of weight 0.85 kg, depending on the engine inlet area, shall not cause the engine to fail suddenly and completely; the engine should continue to deliver thrust over a period of 20 minutes after ingestion.

6.10.1.3   Improved Design and Material Developments of Both Airframe and Engine Parts

- Developments in computer-aided design and manufacturing (CAD/CAM) have led to numerous improvements in both airframe and engine design.
- Carbon-fiber composites, as well as other newly developed materials, have led to superior strength and reduced weight of modules.
- Significant improvements and relocation (within wing and fuselage) is employed for critical components such as fuel lines, flight-control cables, hydraulic lines, and electrical wiring to protect them from bird strikes.
- Development of fly-by-wire control systems have reduced aircraft damage due to bird strike [35].
- Windshields' transparency, strength, flexibility, and bird strike resistance are increased by development in materials and heating systems.
- New composite materials strengthen windshield frames of military aircraft.
- Lighter fighter aircraft became feasible after the adoption of the USAF "Next Generation Transparency Program."

Both FAA and EASA have regulations relating to the design of the different aircraft categories concerning bird strike resistance [35, 36].

Table 6.6 represents the FAA regulations which are extracted from the electronic Code of Federal Regulations [36].

6.10.2   Additional Requirements

As outlined in Aircraft Bird Strike Certification Requirements [37], additional certifications are needed for some categories of aircraft and other aircraft modules.

6.10.2.1   New Aircraft Categories

Bird strike certification requirements are needed for the following:

- Light non-commuter aircraft: this category of aircraft is mostly operating continuously under 8000 ft above mean sea level (AMSL) where almost all bird strikes occur.
- Very light jets: these fly at high speeds. EASA insisted that a new certification item must be added to assure that its windshield must withstand a strike with a bird of a mass at least 0.9 kg (2 lb) at maximum approach flap speed.
- Light turboprop aircraft and helicopters: these are slow-moving aircraft powered by turboprop aircraft and helicopters that have recorded a high proportion of accidents which resulted in damage to the windshield. Numerous accidents justify a review of the bird strike requirements for this category of low-speed light aircraft.

6.10.2.2   Aircraft Modules

Fuel tanks: aircraft fuel tanks have been penetrated in many bird strike cases, resulting in fuel leakage. However, there are no certification requirements for fuel tanks, and this issue needs to be assessed.

Table 6.6 FAA regulations.

Category	Clause	Requirement
Part 23 Normal, Utility, Acrobatic, and Commuter Category Airplanes	23.775(h) (Amdt. 23-49, 11/03/96)	Windshield panes in front of the pilots must withstand, without penetration, the impact of a 2-lb bird when the velocity of the airplane is equal to the airplane's maximum approach flap speed
	23.1323(f) (Amdt. 23-49, 11/03/96)	For commuter category airplanes, pitot tubes must be far enough apart to avoid damage to both tubes in a collision with a bird
Part 25 - Transport Category Airplanes	25.571(e) (Amdt 25-96, 31/03/98)	The airplane must be capable of completing a flight during if impacted with a 4-lb bird when the velocity of the airplane is equal to V_c at sea level or $0.85V_c$ at 8000 ft, whichever is more critical
	25.631 (Amdt 25-23, 08/05/70)	The empennage structure must be designed for safe flight and landing of the airplane after impact with an 8-lb bird when the velocity of the airplane is equal to V_c at sea level
	25.775(b) (Amdt 25-38, 01/02/77)	Windshield panes in front of the pilots and the supporting structures for these panes must withstand, without penetration, the impact of a 4-lb bird when the velocity of the airplane is equal to the value of V_c at sea level
	25.775 (c) Amdt 25-38, 01/02/77	The airplane must have the means to minimize the danger to the pilots from flying windshield fragments due to bird impact
	25.1323(j) Amdt 25-38, 01/02/77	Where duplicate airspeed indicators are required, their respective pitot tubes must be far enough apart to avoid damage to both tubes in a collision with a bird

Electronic flight control systems and flight deck: both FAA and EASA noted that recent electronic flight control systems and flight deck are highly influenced by a bird strike. They are reviewed for possible future certification requirements.

Simultaneous bird ingestion into more than one engine at the same time: this case is not addressed directly, and it is extremely difficult to estimate the probability of its occurrence. One example of it is the Airbus that ditched in the Hudson River in New York. However, since the present standards require that a damaged engine must be shut down, it is hard to apply, especially to a twin-engine aircraft [38].

6.11 Rotary-Wing Aviation

6.11.1 Helicopters

Helicopters normally fly at low altitudes (typically 500 ft AGL) and with low speeds compared to fixed-wing aircraft [32]. Consequently, there is a great possibility for such helicopters to collide with birds. If birds collide with a helicopter's windshield, they may cause serious injury or even incapacity to pilots.

The following points increase the possibility of bird strike:

- the pilot focuses on keeping a minimum terrain clearance during the mission; thus, the pilot has little or no time for watching birds;
- helicopters cause more disturbance to birds than fixed-wing airplanes, and thus, is more susceptible to bird strike risk.

The following suggestions must be adopted to minimize the risk of bird strike with helicopters and the subsequent damage and personnel injury:

- prior to the flight, request bird-activity information;
- fly at higher altitudes;
- review all emergency procedures;
- wear a helmet with a visor;
- if a bird strike occurs, the helicopter should be inspected carefully before the next flight by maintenance engineers;

Concerning certification, helicopters must meet the 14 CFR Part 29-631 standards concerned with structural integrity for large transport types that ensure safe flight and landing after impact by a single bird of mass up to 1 kg at the lesser of V_{ne} and V_h at 8000 ft above mean sea level.

Until now, there is no certification requirement for small helicopters.

6.11.2 Heliports

A heliport is an area of land, water, or structure used for the landing and takeoff of one or more helicopters. It may have limited facilities (supply of fuel or hangars). Other terms used to refer to a heliport are:

- helistop – a minimally developed heliport for boarding and discharging passengers or cargo;
- helipad – is only used in the USA by the FAA, which provides a marked hard surface away from obstacles to provide a safe landing;
- helideck – describes the landing area on a vessel or offshore structure for helicopters landing and taking off.

Heliports must not be in or close to:

- waste disposal facilities such as food-waste landfills;
- waste-transfer stations and compost facilities;
- fish-packing and processing plants;
- abattoirs;
- flight paths of gulls flying between daytime feeding sites and nighttime roosting sites;
- agricultural fields and orchards, harvesting, and plowing activities;
- migratory waterfowl refuges.

6.12   Bird Avoidance

6.12.1   Avian Radar

Researchers observed that birds could be detected by the military radars deployed during the Second World War [38]. By the early 1970s, when inexpensive radars were widely used by marines, biologists, ornithologists, and researchers began using them for studying birds [39]. In the late 1990s, commercially available avian radar systems were introduced to the market by radar companies as a direct response to the growing awareness of the need for affordable avian radar solutions.

However, avian radars are only available in some hub airports. Examples are JFK (in New York City), Seattle-Tacoma, Chicago O'Hare, Dallas–Fort Worth, and Logan/Boston in the USA, Vancouver International Airport in Canada, Schiphol in the Netherlands, Haneda in Japan, Taoyuan in Taiwan, as well as some airports in Turkey and South Africa. Avian radar has also been employed in several military bases.

The main function of avian radar is to detect birds at or close to airports [40]. Radar information is transferred to both the aircraft cockpit and the airport operation center. Thus, pilots can avoid collisions with a large flock of birds.

Radar in its simplest form is the transmission of a pulse of energy, a reflection of a portion of the transmitted energy by a target, and reception of the returned energy by a receiver [41]. In 1940, radar proved its capabilities to detect and track birds. The first major coordinated use of a group of radars to study bird movements over a large region was initiated in Canada in 1964 to address bird collisions with aircraft.

Radar can complement existing management practices (e.g. habitat modification) to reduce the risk of bird collisions with aircraft [12].

6.12.1.1   Avian Radar Fundamentals

Avian radar provides an opportunity to extend observational capabilities to 24/7 time frames and the ability to expand spatial coverage in both distance and altitude [43].

Any avian radar system is composed of four modules, as shown in Figure 6.29. These are:

- the radar unit;
- the scanning unit/antenna;
- the digital radar signal processor (DRSP);
- the visual display.

The radar unit consists of an operational console containing a plan position indicator (PPI) screen. The console is used to adjust settings for power as well as range, antenna rotation speed, and PPI display characteristics.

The avian radar system generates either an electromagnetic energy wave (pulse) or a radio signal, which is next transmitted through an antenna. A part of this signal is reflected from surrounding objects and returns to the system [42]. The returned signal, which is sometimes called an echo, is in an analog form (Figure 6.30). It is first processed into a digital signal, next refined by removing any noise and other interference, and finally, the target is identified and plotted.

The scanning unit (see Figure 6.29) contains, and controls, the system's transmitting and receiving antennas. The antenna generates a directive beam to scan the surrounding

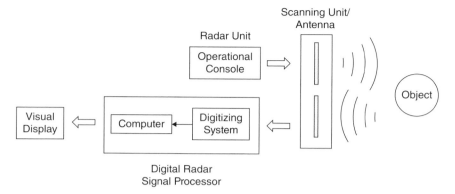

Figure 6.29 Basic radar system components. *Source:* courtesy US Department of Transportation, FAA [42].

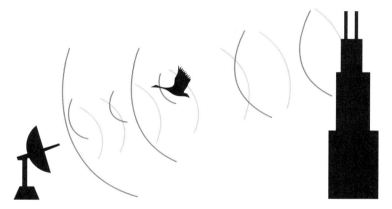

Figure 6.30 Avian radar operation. *Source:* courtesy US Department of Transportation, FAA [42].

area. The antenna is rotated either mechanically or electrically, in either a vertical or a horizontal direction, as needed.

There are two general types of antennas:

- the slotted array antenna (Figure 6.31);
- the parabolic dish antenna (Figure 6.32).

Slotted array antenna have a beam that is 1–2° wide in the horizontal plane and 10° or more in the vertical plane. Thus at 3 miles, such a beam produces coverage that is approximately one-tenth of a mile wide and one-half a mile high. Any target in that volume will be shown on the PPI in the 3-mile range band. Thus, slotted array antennas when spinning in the horizontal plane do not provide any altitude information. Alternately, when spinning in the vertical plane, azimuth and range can be converted to altitude and ground range.

Parabolic dish antennas project a defined conical beam, and thus can provide both range and altitude information. They can rotate at any angle between 0 and 90° above the horizon.

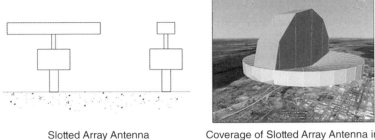

Slotted Array Antenna   Coverage of Slotted Array Antenna in
                                        Vertical and Horizontal Alignment

Figure 6.31 Slotted array antenna. *Source:* courtesy US Department of Transportation, FAA [42].

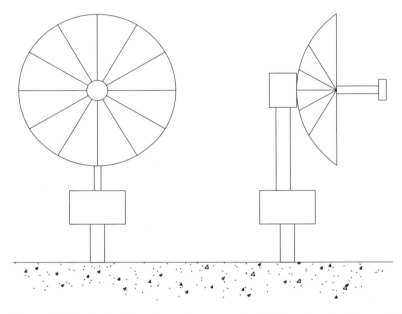

Figure 6.32 Parabolic dish antenna. *Source:* courtesy US Department of Transportation, FAA [42].

The radio signal generated by the radar must produce an echo that can be differentiated from background noise. The echo is characterized by something called the radar cross section (RCS). Figure 6.33 illustrates the measured RCS for a crow.

6.12.1.2   Integration into Airport Operations

Not all airports, but some hub or busy ones, utilize an avian radar system for protecting its aircraft. Data from the radar system are transferred to several airports authorities and departments. However, airport operators must be sure that avian radar systems do not interfere with ATC equipment.

Avian radar systems are capable of:

- day and night monitoring of any bird movements;
- permanent automatic recording of all targets detected and tracked;

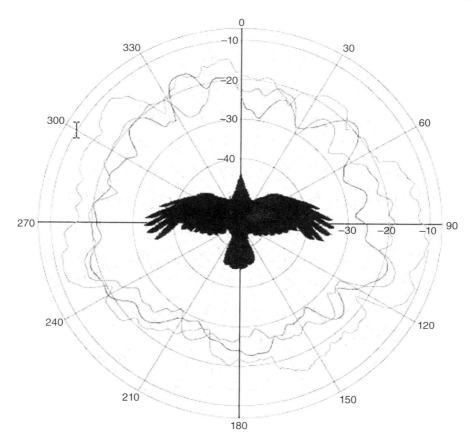

Figure 6.33 The measured RCS for a crow. *Source:* courtesy US Department of Transportation, FAA [42].

- developing hourly, daily, weekly, monthly, and seasonal summaries of bird activities;
- extending surveillance areas (both in distance and altitude).

    Avian radar systems yield the following short-term benefits to airports:

- they transmit radar images to airport authorities and thus help managers to have quick responses to bird threats;
- they identify any sudden bird hazards that are not detected by traditional bird hazard management methods
- they add any unusual bird activities to the update of information in airport ATIS announcements or NOTAMs.

    The airport's long-term benefits from radar data analysis include:

- developing an archive for information to be used in the airport or shared with other airports (based on 37 AC 150/5220-25 11/23/10);
- identifying mitigation activities (times and routes);
- identifying the status of transitory birds (origin and destination);

- identifying the relationships between habitat characteristics and bird activity at hazardous areas;
- understanding the characters of bird movements and abundance;
- studying the bird management efforts.

A dual radar configuration with one antenna sweeping vertical and the second sweeping horizontal [44] will provide three-dimensional coverage for critical runways and approach and departure corridors (Figure 6.34). It can function from ground level up to 15 000 ft AGL and 360° around the airfield out to 8 or more miles. It can track birds, even in fog, rain, and snow.

Avian radar may be fixed or small mobile devices (Figure 6.35). The famous types of radar are Robin, Merlin, and Accipiter.

In January 2010, Seattle-Tacoma International Airport was the first civil airport to deploy real-time bird tracking, using the Accipiter Avian Radar system. It can detect individual birds ranging from small sparrows to large Canada geese up to 2 miles away.

Figure 6.34 Dual radar configuration. *Source:* courtesy US Department of Transportation, FAA [42].

Figure 6.35 A mobile avian radar. *Source:* courtesy USDA-APHIS [48].

Other airports currently using other types of avian radar are Schiphol Amsterdam (four units), Frankfurt (one unit), Berlin Brandenburg (one unit), Hatay, Turkey (two units), the Dutch Airforce, including Eindhoven airport (four units).

Robin Radar has been a technology leader in the field. It has developed technology both in hardware and software.

Robin Radar launched the third generation of avian radar systems in 2017. This system is called MAX and has unprecedented performance levels. The MAX radar is the first purpose-built full 3D bird radar (Figures 6.36–6.38). It enables the detection of birds up to 10 km away at 1 km altitude. 3D track updates are real time, so every bird (SAT1) within 10 km has longitude/latitude/altitude positioning in real time.

Figure 6.36 Robin radar. *Source*: courtesy Robin Rada Systems [57].

Figure 6.37 Robin Max Radar. Source: courtesy Robin Rada Systems [57].

Figure 6.38 Robin Max Radar close to runways. Source: courtesy Robin Rada Systems [57].

The main features of MAX are:

- single sensor, phased array (FMCW), stacked beam radar, X-band;
- limited cone of silence (only 30°);
- detection capabilities: SAT 1: 10–13 km range / 1 km altitude;
- detection capabilities: song bird: 5–6 km range / 500 m altitude.
- track update speed: every second (real time).

The system is currently in operation at Copenhagen airport in Denmark (since March 2018). The Hague airport will have a MAX system operational by December 2018. Moreover, by the first quarter of 2019, the Dutch Airforce is going to apply two of these

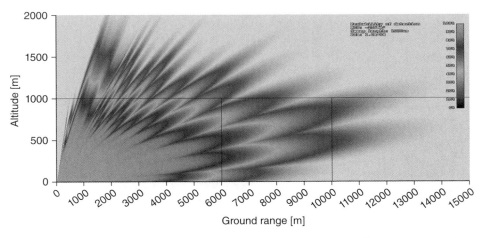

Figure 6.39 The detection range for a SAT 1 (small goose target) at 16 dB m². Source: courtesy Robin Rada Systems [57].

systems at their prime fighter bases (Volkel and Leeuwarden AB), as well as one system to their bombing range at Cornfield-Vliehors (Rotterdam).

Figure 6.39 illustrates the detection range for a SAT 1 target (such as a small goose) at 16 dB m².

Figure 6.40 illustrates the detection graph for a songbird at 25 dB m².

Since large airports typically need more than one radar for a full coverage requirement, several radar nodes are employed, as shown in Figure 6.41, with a break-out of one of them. Each node resembles an avian radar consisting of a radar sensor/transceiver (RST) mounted on a platform, a digital radar processor (DRP), and a remote radar controller (RRC). The RST includes at least one scanner/antenna and corresponding transceiver(s). The RRC and DRP are connected over a network via (TCP/IP) Transmission Control

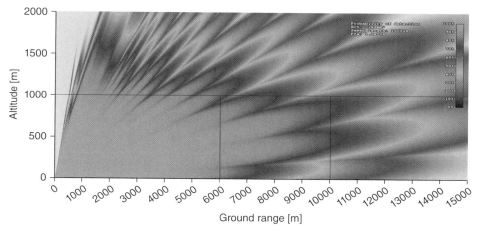

Figure 6.40 The radar detection graph for a songbird at 25 dB m². Source: courtesy Robin Rada Systems [57].

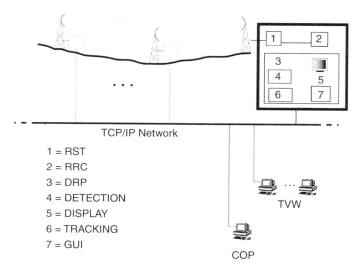

1 = RST
2 = RRC
3 = DRP
4 = DETECTION
5 = DISPLAY
6 = TRACKING
7 = GUI

Figure 6.41 3D avian radar systems design.

Protocol/Internet Protocol. The RRC provides a network interface to the RST. Thus, it could be turned on/off and configured remotely by a radar technician. The DRP is also connected to a TCP/IP network for remote operation and remote video display (for the radar technician), and for target data distribution.

6.12.2 Optical Systems

Optical systems includes video, image intensification, and thermal systems. Video cameras are sensors that capture images. They provides not only target detection but also

Figure 6.42 Airport camera installation. *Source:* courtesy FAA [49] and [50].

Figure 6.43 Multiple cameras monitoring birds on runways. *Source:* courtesy FAA [49] and [50].

support target identification. Cameras are mounted on rigid towers [49, 50]. Examples are available in the following airports: Chicago O'Hare (USA), Heathrow (UK), and Vancouver (Canada) (Figure 6.42).

Multiple cameras may also be fixed along runways, as installed in Chicago O'Hare (Figure 6.43).

Magnification can reveal the detail needed for identification [49, 52] (Figure 6.44).

An infrared camera uses an illuminator to capture images (Figure 6.45).

A combination of radar and a set of cameras is used to monitor bird movements close to [44] and far away.

Figure 6.44 Magnification of single bird. *Source:* courtesy FAA [49] and [51].

Figure 6.45 Bird identification at night using an infrared camera. *Source:* courtesy FAA [49] and [50].

References

1 Cleary, E.C. and Dolbeer, R.A. (2005). *Wildlife Hazard Management at Airport: A Manual of Airport Personnel*, 2e, 1–135.

2 Morris, H. (2017). How many planes are there in the world right now? *The Telegraph*, 16 August 2017.

3 DeVault, T.L., Blackwell, B.F., Belant, J.L., and Begier, M.J. (2017) Wildlife at Airports: Wildlife Damage Management. Technical Series, US Department of Agriculture Animal & Plant Health Inspection Service Wildlife Services.

4 Bird/Wildlife Aircraft Strike Hazard (BASH) Management Techniques, Air Force Pamphlet 91–212. 1 February 2004.

5 Airbus Industries, Birdstrike Threat Awareness. Flight Operation Briefing Notes, Operating Environment.

6 Nicholson, R. and Reed, W.S. (2011). Strategies for Prevention of Bird-Strike Events, *AERO* QTR_03.11, Boeing. http://www.boeing.com/commercial/aeromagazine/ articles/2011_q3/4/ (accessed 2 January 2019).

7 Aircraft Damages Caused by Birds, Aviation Ornithology Group, Moscow. http://www .otpugivanie.narod.ru/damage/eng.html (accessed 29 December 2018).

8 Maragakis, I. (2009). European Aviation Safety Agency, Safety Analysis and Research Department Executive Directorate, Bird Population Trends and Their Impact on Aviation Safety 1999–2008.

9 FAA Wildlife Strike Database. http://wildlife.faa.gov (accessed 9 January 2019).

10 Search the FAA Wildlife Strike Database. http://wildlife.faa.gov/database.aspx (accessed 9 January 2019).

11 Airbus Flight Operations Briefing Notes; Operating Environment: Birdstrike Threat Awareness. French DGAC.

12 Blokpoel, H. (1974). *Bird Hazards to Aircraft*, 236. Ottawa, Ontario, Canada: Canadian Wildlife Service. Ministry of Supply and Services.

13 International Birdstrike Committee (IBSC) (2006). Recommended Practices No. 1: Standards For Aerodrome Bird/Wildlife Control, Issue 1. http://www.int-birdstrike. org/Standards_for_Aerodrome_bird_wildlife%20control.pdf (accessed 9 January 2019).

14 Bishop, J., McKay, H., Parrott, D. and Allan, J. (2003). Review of international research literature regarding the effectiveness of auditory bird scaring techniques and potential alternatives. https://www.researchgate.net/publication/242454383_ Review_of_international_research_literature_regarding_the_effectiveness_of_ auditory_bird_scaring_techniques_and_potential_alternatives (accessed 9 January 2019).

15 Murton, R.K. and Wright, E.N. (eds.) (1968). *The Problems of Birds as Pests*, Symposia of the Institute of Biology No. 17. Academic Press.

16 Tina Turner scares birds at Gloucestershire Airport, ITV News. https://www.itv.com/ news/westcountry/update/2012-11-03/tina-turner-scares-birds-at-gloucestershire- airport/ (accessed 15 January 2019).

17 Blackwell, B.F., Bernhardt, G.E., and Dolbeer, R.A. (2002). Lasers as non-lethal avian repellents. *Journal of Wildlife Management* 66 (1): 250–258.

18 Battistoni, V., Montemaggiori, A., and Iori, P. (2008). Beyond falconry between tradition and modernity: a new device for bird strike hazard prevention at airports.

In: *Proceedings of International Bird Strike Committee, IBSC Meeting, and Seminario Internacional Perigo Aviario e Fauna*, 1–13. Brasilia.

19 Watermann, U. 1997. Experimental falconry program to reduce the gull strike hazard to aircraft at John F. Kennedy International Airport, New York. Report prepared for Port Authority of New York and New Jersey by Bird Control International Inc., Georgetown, Ontario, Canada.

20 Precise Flight. https://www.preciseflight.com/commercial (accessed 9 January 2019).

21 Harris, R.E. and Davis, R.A. (1998). Evaluation of the efficacy of products and techniques for Airport Bird Control. LGL Report TA2193. LGL Limited, Environmental Research Associates.

22 O'Callaghan J. (2014). Is it a bird? Is it a plane? No, it's ROBIRD: Robotic falcons and eagles mimic predators to keep pests away from airports and farms. www.dailymail. co.uk/sciencetech/article-2743272/Is-bird-Is-plane-No-s-ROBIRD-Robotic-falcons-eagles-mimic-real-predators-pests-away-airports-farms.html#ixzz4X4HvwaLs (accessed 9 January 2019).

23 Dolbeer, R.A., Chipman, R.B., Gosser, A.L. and Barras, S.C. (2003). Does Shooting Alter Flight Patterns of Gulls: A Case Study at John F. Kennedy International Airport. International Bird Strike Committee, IBSC 26/WP-BB5, Warsaw. http://www.int-birdstrike.org/Warsaw_Papers/IBSC26%20WPBB5.pdf (accessed 9 January 2019).

24 Dolbeer, R.A. (1998). Evaluation of Shooting and Falconry to Reduce Bird Strikes with Aircraft at John F. Kennedy International Airport. International Bird Strike Committee, IBSC 24/WP 13, Stara Lesna, Slovakia. http://citeseerx.ist.psu.edu/viewdoc/download; jsessionid=0A84081FDCF9D113CB9A002A648F4987?doi=10.1.1.567.6216&rep=rep1 &type=pdf (accessed 9 January 2019).

25 DeFusco, R.P. and Unangst, E.T. (2013**).** Airport Wildlife Population Management: A Synthesis of Airport Practice. Airport Cooperative Research Program, ACRP Synthesis Volume 39.

26 Desoky, A.E.A.S.S. (2014). A review of bird control methods at airports. *Global Journal of Science Frontier Research: E Interdisciplinary* 14 (2).

27 Hygnstrom, S.C., Timm, R.M., and Larson, G.E. (eds.) (1994). *Prevention and Control of Wildlife Damage*. Lincoln, Nebraska: University of Nebraska Cooperative Extension Division.

28 Dolbeer, R.A., Seamans, T.W., Blackwell, B.F., and Belant, J.L. (1998). Anthraquinone formulation (Flight Control™) shows promise as avian feeding repellent. *The Journal of Wildlife Management* 62 (4): 1558–1564.

29 Transport Canada (1994), TP11500 E). *Wildlife Control: Procedures Manual*. Safety and Security Aerodrome Safety Branch http://publications.gc.ca/collections/collection_2013/tc/T52-4-79-2002-eng.pdf (accessed 9 January 2019).

30 Matyjasiak, P., Methods of bird control at airports, Theoretical and Applied Aspects of Modern Ecology, J. Uchmański (ed.), Cardinal Stefan Wyszyński University Press , Warsaw, pp. 171–203, 2008, https://www.researchgate.net/publication/233389769_Methods_of_bird_control_at_airports (accessed 9 January 2019).

31 Belant, J.L. (1997). Gulls in urban environments: landscape-level management to reduce conflict. *Landscape and Urban Planning* 38: 245–258.

32 Sharing the Skies: An Aviation Industry Guide to the Management of Wildlife Hazards (Transport Canada), 22004. https://www.ascendxyz.com/wp-content/uploads/regulatory-requirements/Reference%208_Sharing%20the%20Skies.%20An

%20Aviation%20Industry%20Guide%20to%20the%20management%20of
%20Wildlife%20Hazards.%20Transport%20Canada.pdf (accessed 9 January 2019).

33 Airbus (2006). Flight Operations Briefing Notes, Supplementary Techniques, Handling Engine Malfunctions. http://www.smartcockpit.com/docs/Handling_Engine_Malfunctions.pdf (accessed 9 January 2019).

34 Aircraft Certification for Bird Strike Risk. http://www.skybrary.aero/index.php/Aircraft_Certification_for_Bird_Strike_Risk (accessed 9 January 2019).

35 Bird Strike Damage & Windshield Bird Strike, ATKINS Final Report, 5078609-rep-03, Version 1.1. https://www.easa.europa.eu/system/files/dfu/Final%20report%20Bird%20Strike%20Study.pdf (accessed 29 December 2018).

36 Electronic Code of Federal Regulations. https://gov.ecfr.io/cgi-bin/text-idx?SID=bc22e7545c0fd0578edbe95e59cd1e21&mc=true&tpl=/ecfrbrowse/Title23/23cfrv1_02.tpl#0 (accessed 15 January 2019).

37 Airport Bird Hazard Management. https://www.skybrary.aero/index.php/Airport_Bird_Hazard_Management (accessed 10 January 2019).

38 Aircraft Certification for Bird Strike Risk. https://www.skybrary.aero/index.php/Aircraft_Certification_for_Bird_Strike_Risk (accessed 10 January 2019).

39 Nohara, T.J., Weber, P., Ukraine, A. and Premji, A. (2007). An Overview of Avian Radar Developments – Past, Present and Future, Bird Strike Committee Proceedings, 9th Annual Meeting, Kingston, Ontario, September 10–13, 2007. http://digitalcommons.unl.edu/cgi/viewcontent.cgi?article=1012&context=birdstrike2007 (accessed 10 January 2019).

40 Air Line Pilots Association, International (2009). Wildlife Hazard Mitigation Strategies for Pilots. https://www.alpa.org/-/media/ALPA/Files/pdfs/news-events/white-papers/wildlife-hazard.pdf?la=en (accessed 10 January 2019).

41 Gauthreaux, S.A. Jr. and Schmidt, P.M. (2013). Wildlife in airport environments: Chapter 13 RADAR technology to monitor hazardous birds at airports. In: *Wildlife in Airport Environments: Preventing Animal-Aircraft Collisions through Science-Based Management* (ed. T.L. DeVault, B.F. Blackwell and J.L. Belant). Johns Hopkins University Press.

42 US Department of Transportation (2010). Airport Avian Radar Systems, Federal Aviation Administration, Advisory Circular, AC No: 150/5220-25. https://www.faa.gov/documentLibrary/media/Advisory_Circular/AC_150_5220-25.pdf (accessed 10 January 2019).

43 Brand, M., Key, G., Nohara, T.J., et al. (2011). Integration and Validation of Avian Radars (IVAR), Final Report, ESTCP Project RC-200723. https://apps.dtic.mil/dtic/tr/fulltext/u2/a554410.pdf (accessed 15 January 2019).

44 Bird Detection System. NEC Corporation, Air Transportation Solutions Division, Tokyo, Japan. https://www.nec.com/en/global/solutions/bird/common/pdf/bird_detection_system.pdf (accessed 10 January 2019).

45 International Birdstrike Committee (2016). Recommended Practices No. 1: Standards For Aerodrome Bird/Wildlife Control. http://www.int-birdstrike.org/Standards_for_Aerodrome_bird_wildlife%20control.pdf (accessed 10 January 2019).

46 Do All Dogs Like to Chase Birds? Cuteness Team. https://www.cuteness.com/blog/content/do-all-dogs-like-to-chase-birds (accessed 10 January 2019).

47 Vultures. Wildlife Damage Management Technical Series, US Department of Agriculture, Animal and Plant Health Inspection Service. https://www.aphis.usda.gov/aphis/ourfocus/wildlifedamage/operational-activities/SA_Vultures (accessed 10 January 2019).

48 Avian Radar: Does it Work? United States Department of Agriculture, Animal and Plant Health Inspection Service. https://www.aphis.usda.gov/aphis/ourfocus/wildlifedamage/programs/nwrc/sa_spotlight/avian+radar+does+it+work (accessed 10 January 2019).

49 Weller, J.R. (2014). FOD Detection System: Evaluation, Performance Assessment and Regulatory Guidance. Wildlife and Foreign Object Debris (FOD) Workshop, Cairo, Egypt, March 24–26, 2014. https://www.icao.int/MID/Documents/2014/Wildlife%20and%20FOD%20Workshop/Assessing%20Risk%20FAA.pdf (accessed 10 January 2019).

50 Wildlife Strikes to Civil Aircraft in the United States 1990–2014. FAA National Wildlife Strike Database: Serial Report Number 21. http://www.fwspubs.org/doi/suppl/10.3996/022017-JFWM-019/suppl_file/10.3996022017-jfwm-019.s8.pdf (accessed 10 January 2019).

51 Heathrow Imagery. http://www.tarsier.qinetiq.com/imagery/Pages/default.aspx#!prettyPhoto/2 (accessed 10 January 2019).

52 Tarsier. http://www.tarsier.qinetiq.com/solution/Pages/tarsier.aspx#!prettyPhoto/4 (accessed 10 January 2019).

53 Robop Ltd. https://robop.com/brochure/index.html (accessed 10 January 2019).

54 Eagle Helicopter. https://www.snopes.com/fact-check/awesome-paint-job/#photo (accessed 15 January 2019).

55 Wind Powered Flashing Hawkeye Bird Scarer. https://www.scaringbirds.com/wind-powered-scarers/flashing-hawkeye-ground-mounted (accessed 10 January 2019).

56 Bird Strike https://en.wikipedia.org/wiki/Bird_strike (accessed 10 January 2019).

57 Private communications, Robin Rada Systems.

7

Airframe and Engine Bird Strike Testing

7.1   Introduction

Aircraft may encounter a bird strike. Such strikes will not be only a serious encounter for the birds; it is also a major risk for those inside the aircraft. A bird strike may result in:

- aircraft damage or crash that costs millions of dollars;
- damage to aircraft property;
- death or injury to humans.

Bird strike resistance is now a necessary issue for certification of both airframe and engine parts, as set by American, European, and Canadian authorities: the FAA (Federal Aviation Agency), the European Aviation Safety Authority (EASA), and Transport Canada (TC).

Aircraft parts must pass a series of bench tests, where the part is mounted on a rig and struck by a bird fired at a realistic operational velocity from a high-powered gas cannon.

Bird Strike in Aviation: Statistics, Analysis and Management, First Edition. Ahmed F. El-Sayed.
© 2019 John Wiley & Sons Ltd. Published 2019 by John Wiley & Sons Ltd.

Numerical methods assist in minimizing the number of experimental test cases and thus save a lot of the testing costs. Numerical methods have been adopted since the 1980s to optimize structural design against bird strike and guarantees that the first prototypes used in certification tests provide satisfactory results. Chapter 8 will describe in detail various numerical techniques employed in bird strike simulation.

This chapter discusses the experimental methods that simulate an airborne bird strike and the simulate the resulting damage to both airframe and engine components. Tests are performed according to the Standard Test Method [1].

All exterior parts of an aircraft having a forward-facing projected area are subject to bird strike. The tested parts should be representative of the actual hardware and the supporting structure of the aircraft, in order to take into account the real dynamic response of the part to the actual bird strike event [2]. The testing should include extreme environmental conditions that may be encountered in an actual bird strike. As identified by Speelman et al. [2], testing should include impact locations of:

- maximum stiffness;
- maximum deflection;
- the critical support structure and actuating mechanisms;
- power and fuel lines, hydraulic circuits;
- impact shock dynamics activating/dislodging electrohydraulic switching;
- actuating mechanisms critical for continued flight.

7.2 Bird Impact Test Facilities

7.2.1 Introduction

The main components of any bird impact test facility [2] are:

- pressurized air tank;
- pressure release valve;
- chamber for holding a sabot that holds the bird (impact projectile);
- tube for accelerating the bird (projectile);
- instrumentation for measuring both velocity and orientation of the projectile;
- station for mounting the item to be impacted;
- the backstop for absorbing residual energies;
- electrical interconnections for safety and data acquisition;
- insulating blankets or curtains for either heating or cooling equipment in the impact area;
- high-speed photography;
- additional control for automatic firing of birds in case of testing rotating blades of jet engines, to assure hitting the desired location.

The minimum capture rate of high-speed photography is 5000 frames per second (fps) for accurate analysis of results. Multiple cameras and lighting are synchronized as a part of the automatic firing sequence.

Some remarks may be added concerning rotating element testing:

- a precise launching sequence is conducted such that the bird goes between two blades and the back of the blade hits the bird;

- the rotating item is connected to the drive mechanism through frangible couplings;
- In some cases multiple launch tubes are used and spring-loaded mechanisms are used instead of the air cannon to launch the projectiles.

These testing facilities simulate bird strikes at extremely high speeds, i.e. speeds greater than $900\,km\,h^{-1}$ (which is equivalent to Mach 0.75).

7.2.2 Test Facilities

7.2.2.1 USA
Bird strike facilities in the USA are found in the following locations.

A. For non-rotating parts:
  - Arnold AFB, TN 37389;
  - Boeing Commercial Aircraft Co., Seattle, WA 98124;
  - Lockheed-Martin Tactical Aircraft Systems, Worth, TX 76101-0748;
  - PPG Industries, Inc., Huntsville, AL 35804;
  - University of Dayton Research Institute, Dayton, OH 45469-0101.
B. For rotating parts:
  - Air Force Research Laboratory's Bird Impact Test Range;
  - Wright-Patterson AFB, OH, 45433-6563.
  - Each engine manufacturer, such as General Electric and Pratt & Whitney, also has its facility for use for their engines.

7.2.2.2 Canada
Bird strike facilities are available at:

  - National Research Council of Canada (NRC);
  - Pratt & Whitney Canada;
  - University of New Brunswick;
  - Bombardier.

7.2.2.3 Europe
The main bird strike test facilities are found at:

  - Rolls Royce (UK);
  - Dassault Aviation;
  - Airbus Industries;
  - University of Ghent (Belgium);
  - Alena plant (Italy);
  - Lufthansa Technik (Germany).

7.3 Details of Some Test Facilities

Bird strike testing started in the 1950s when the chicken gun (a large diameter compressed air cannon) was invented by de Havilland in the UK. Freshly killed chickens were packed into a compressed air gun and fired at the aircraft module to be tested (e.g. windscreen and engines).

Boeing used fowl to test aircraft structures in its Flight Test Division.

7.3.1 Aircraft Windshield and Airframe Testing

Aircraft windshields and airframes are tested using compressed air cannons. They direct euthanized birds against different airframe components at specified speeds. The following parameters are measured: the bird's speed and acceleration, the pressure, displacement, vibration and impact of the strike, and component distortion (the amount of stretch and deformation of the materials). Moreover, ultra-high-speed cameras and data acquisition systems are used to identify the progression of the damage. Bird strike testing is performed according to the Standard Test Method [1].

7.3.1.1 Chicken Gun or Chicken Cannon

A chicken gun was designed to simulate accurately a large, live bird in flight. It is used to test the strength of critical aircraft modules, such as its windshields, as well as the fan blades of jet engines.

It is important to notice that:

- cockpit windows are necessarily made of thin transparent materials, and thus are a vulnerable spot;
- fan blades and other rotating blades in jet engines are subjected to serious damage, which may lead to catastrophic accidents.

As displayed in Figure 7.1, the cannon is used to fire the chicken into the engine, windshield, or other test structure.

The gun is driven from a compressed-air tank. To fire the gun a solenoid-driven needle strikes and rupture the diaphragm, allowing the compressed air to drive the chicken, surrounded by a cylindrical cardboard ice cream carton, down the barrel. At the muzzle, a metal ring stops the carton but allows the chicken to pass through. Slow-motion cameras photograph the chicken impacting the target in the test bed. These cameras start at the same time as the breaking of the diaphragm.

7.3.1.2 Alenia Plant Testing

The Alenia plant test facility [3] is illustrated in Figure 7.2. An air pressure gun is used to shoot the bird at the test target at the desired speeds. This apparatus has a barrel length of 12 m, and it is capable of generating impact speeds up to $140 \, \mathrm{m \, s^{-1}}$ (313 mph).

7.3.2 Engine Testing

The first rotating element of a turbofan engine is its fan while that for a turbojet engine is its compressor.

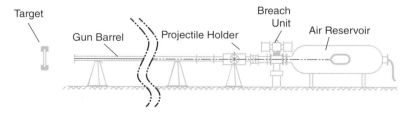

Figure 7.1 A typical chicken gun.

Figure 7.2 An air pressure gun. *Source:* Reproduced with permission [3].

Such rotating turbomachinery is subjected to impact-loading tests to verify their structural integrity. Artificial or euthanized birds are fired into running engines from multi-barrel air guns.

Different engine parameters, such as like pressures, temperatures, accelerometer forces, and strain-gauge values, are recorded both during the impact and for a specified time after the impact. Impact tests are also photographed using high-speed cameras to trace bird trajectories and observe engine-blade deformation.

Figure 7.3 shows the engine bird-impact test facilities at Rolls Royce.

7.3.3 Artificial Birds Versus Real Birds

The different types of birds used in testing are either artificial or real (wild or domestic) birds.

Figure 7.3 Aircraft engine bird-impact testing. *Source:* Courtesy Rolls-Royce [4].

The choice of bird type for testing depends on several factors [2].

Many airframe/aero-engine companies use artificial birds or substitute material in its pre-certification experimental tests. However, final certification testing is normally carried out with real birds, which is both costly and time consuming.

7.3.3.1 Real Birds

Real birds used in testing are either wild or domestic. The following remarks are for real bird testing.

- Use of real bird bodies in aircraft component testing is not ideal.
- Wild birds are costly to acquire, and environmental protection considerations make it difficult to justify their use.
- Examples of real birds used in testing are Canada geese and chickens (pre-euthanized), weights are 2.5, 4, 6, and 8 lb).
- Domestic birds such as chickens are bred to have a different structure than wild birds.
- All real birds used in testing are painlessly killed, frozen/refrigerated until ready for the test, and then warmed to room temperature and adjusted in weight to the desired test condition.
- Wild birds "spread out" much more than chickens when shot out of a bird gun; as a result, much of the wild bird's mass will miss the "weak point" on the aircraft.
- Both wild birds and domestic birds are costly to use when facility clean-up after each test is considered.
- Tests using real birds are not uniform as authorities define only the masses to be used and not the species. Variations in bird body density between species and even between individuals of the same species may cause different and unpredictable effects upon impact [5].

7.3.3.2 Artificial Birds

Artificial birds have been used widely in the aerospace industry to counter many of the problems described above with real birds. Tests using artificial birds have the following features:

- Tests using artificial birds create realistic impact loading under the proper conditions.
- Artificial birds are economical both in preparation and the clean-up stages.
- Artificial birds may be fabricated of a gelatin material with 10% porosity having a nominal density of 0.92 g/cm^3.
- Artificial birds made from gelatin may have any of the following simplified geometries (Figure 7.4):
  - circular cylinder with straight ends;
  - circular cylinder with hemispherical ends;
  - ellipsoid.
- The shapes of artificial birds are selected to permit the most appropriate values of mass, density, diameter, and length for a projectile [5].
- It may be appropriate to use an accurate value for the length of the projectile (impact time histories for impact onto a flat panel). Alternatively, the diameter may be the most critical parameter (for instance when testing the slicing effect of rotating blades in turbomachinery).

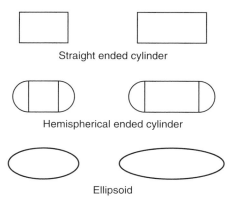

Straight ended cylinder

Hemispherical ended cylinder

Ellipsoid

Figure 7.4 Three suggested artificial bird shapes, shown to scale.

7.4 Certification Requirements

Canadian, European, and American authorities (TC, EASA, and FAA) have their own regulations regarding certification requirements for both airframe and engines [6, 7]. The following requirements must be fulfilled:

a. Both non-commuter light aircraft (CS-23) and light rotorcraft (CS-27) have no bird strike certification requirements.
b. EASA insisted that very light jets (VLJs) must have the same certification requirements as commuter light aircraft (CS-23) due to their high speed; i.e. the windshield should be able to withstand a strike with a bird of a mass at least 0.9 kg (2 lb) at maximum approach flap speed.
c. Based on FAR 25 (section 775), the airframe certification criteria [3] of larger aircraft necessitates that the aircraft should be able to safely continue flying after striking a 1.8 kg (4 lb) bird at design cruise speed (V_c).
d. Concerning aircraft empennage, the new requirement necessitates that it must withstand impacts of 3.6 kg (8 lb), based on FAR 25 (section 571).
e. There are no specific bird strike certification standards for fuel tank areas. They must withstand impacts of 1.8 kg, as all other aircraft modules.

7.5 Airframe Testing of Transport Aircraft

Some details about the testing of wing and empennage of transport aircraft will be discussed here. Concerning military aircraft, a few details concerning tests of the canopy and the windscreen of the F35, as well as the inlet door of the lift fan of the F35 in its STOVL (short take-off and vertical landing), are also outlined.

7.5.1 Wing Testing

Currently, commercial aircraft operate at high speeds that approach $1000 \, \text{km h}^{-1}$. As a result, the design, construction, and testing of aircraft wings must match with the requirements for bird strike certification standards. If a bird strikes the wing, testing must guarantee that the aircraft will be able to land safely.

7.5.1.1 Case Study

The Alenia C-27J Spartan military transport aircraft (developed and manufactured by Leonardo's Aircraft Division) was examined. Testing of the wing leading edge structure of the Alenia C-27J was carried out at the Alenia plant, Italy [3]. A gas gun facility is used for testing a one-bay component of its wing leading edge. A parametric study using several new configurations, various materials, lay-up distribution, and boundary conditions was performed. The objective of the test was to check whether new configurations of leading edge structure will be able to satisfy the bird strike requirement according to the Federal Aviation Regulation (FAR part 25, section 25.631 "Bird-strike Damage") or not.

The test facilities are illustrated in Figures 7.5 and 7.6 and have the following features:

- the whole assembly was located 3 m from the mouth of the gun;
- the barrel length of the air pressure gun is 12 m;
- typical dimensions of tested wing section are 640 mm × 330 mm and thickness of the rib is 2 mm;
- the projectile (a dummy bird) is adjusted to impact the wing leading edge at its mid height;
- the projectile was held inside a sabot packed with expanded polystyrene such that no position changes or damage would be experienced under the acceleration when fired;
- dummy birds are shot at speed up to 140 m s^{-1};
- the support frame was suspended using six load cells of 10 kN to measure loads transmitted to the foundations in three dimensions;
- the system was linked to the loading frame and several sensors (accelerometers, strain gauges) around the test specimen;

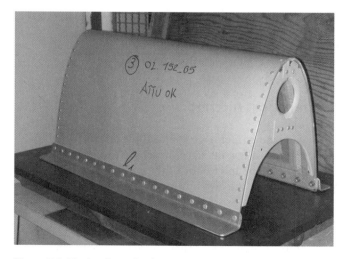

Figure 7.5 The leading edge bay. *Source:* Reproduced with permission [3].

Figure 7.6 Testing rig. *Source:* Reproduced with permission [3].

- two high-speed cameras, recording up to 10000 fps, are used to visualize and record the impact sequence in the front and the lateral views;
- a data acquisition system having a frequency of 100 kHz is used.

The experimental results are shown in Figure 7.7.

7.5.2 Empennage Testing

The EASA Certification Specifications, CS 25 (section 631) [8], states that:

> The empennage structure must be designed to assure capability of continued safe flight and landing of the airplane after impact with an 8 lb bird when the velocity of the airplane (relative to the bird along the airplane's flight path) is equal to cruise velocity (V_c) at sea level.

7.5.2.1 Case Study 1

The leading edge of the vertical empennage of a cargo airplanes Alenia C-27J was tested for certification of bird strike resistance. The leading edge of the vertical empennage

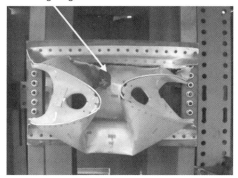

Figure 7.7 Deformed shape of the front and rear view. *Source:* Reproduced with permission [3].

has a total length of 2.970 m, with a chord that varies between 0.450 and 0.750 m (Figure 7.8). This tested fin was connected to a test rig attachment (Figure 7.9).

Figure 7.10 illustrates the shape of the leading edge after the bird strike at 250 knots. The strike produced a large deformation, but the test is considered passed because the test article has survived.

7.5.2.2 Case Study 2

The bird strike tests of the new Russian MS-21 airliner have been carried out [9] at the Central Institute of Aviation Motors (CIAM) on 6 August 2015. The jet was tested using a pneumatic cannon. The aircraft was fixed, and a bird's carcass with a weight of 1.8 kg

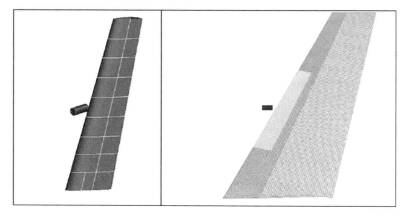

Figure 7.8 General view of the leading edge and its assembly on the fin of Alenia C-27J. *Source:* Reproduced with permission [3].

Figure 7.9 Test rig attachment. *Source:* Reproduced with permission [3].

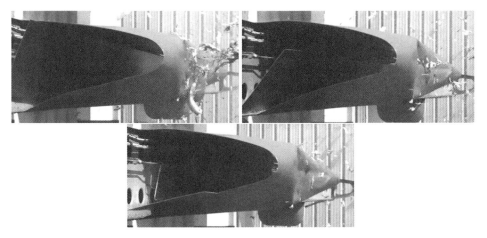

Figure 7.10 Experimental correlation at different time steps. *Source:* Reproduced with permission [3].

was launched using the pneumatic cannon with speed equal to the aircraft's flight speed. According to experts, the airliner's fin was certified.

7.6 Airframe Testing of Military Aircraft

Critical areas in an F35 are the canopy, the windscreen, the lift fan inlet door, the auxiliary air intake and doors, and the engine inlet structure (Figure 7.11). Analyses for bird strike resistance of the canopy, the windscreen as well as the inlet door of the lift fan of F35B in its STOVL form will be described here [11].

Figure 7.11 F35 [10].

7.6.1 Canopy and Windscreen

The canopy system must withstand the impact of a 4 lb bird at 480 knots on the rein-forced windscreen and 350 knots on the canopy crown without:

- breakage/deflection that strikes the pilot when seated in the design eye "high" position;
- damage to the canopy that may result in incapacitating injury to the pilot;
- damage that would preclude safe operation of, or emergency egress from, the aircraft.

Figure 7.12 illustrates the bird impact test facility and setup at the Lockheed Martin company [11].

The test locations for the windscreen and canopy are illustrated in Figure 7.13:

- windscreen Tests – "high" and "low" shots at 480 knots with a 4 lb bird;
- canopies shots at 350 knots with a 4 lb bird.

The impact locations are shown in Figure 7.14 [11].

Figure 7.12 Bird strike testing [11].

Figure 7.13 Tested areas on F35B [11].

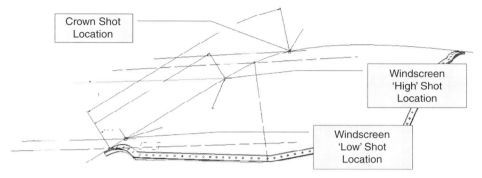

Figure 7.14 Locations of shots on windscreen and canopy [11].

The sequence of impacts of a 4 lb chicken shot at 480 knots at the canopy are shown in Figure 7.15. The requirements for canopy bird impact design were successfully verified by tests.

7.6.2 Lift Fan Inlet Door (STOVL Mode)

The lift fan inlet door was selected for testing due to concern that bird or door structural debris could be ingested into the lift fan or become foreign object damage (FOD) for the main engine via the open auxiliary air intake (Figure 7.16).

Figure 7.17 illustrates the impact location with the lift fan inlet door of a F35B. The door angle was set at the "full open" position plus increment to account for aircraft angle of attack. Impact location was centered on the largest bay over the fan. Tests for score used a 2 lb bird shot at 140 knots. The lift fan inlet door passed this bird strike test.

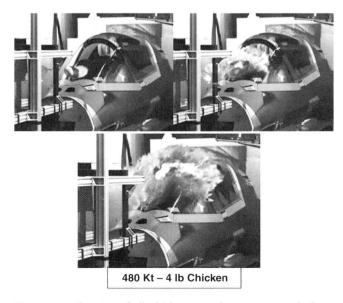

480 Kt – 4 lb Chicken

Figure 7.15 Shooting of 4 lb chicken at 480 knots at canopy [11].

Figure 7.16 F35B with lift fan inlet door and auxiliary air doors open [11].

Figure 7.17 Impact testing location with lift fan inlet door of F35B [11].

7.7 Engine Testing of Civil and Military Aircraft

The jet engines that power nearly all airliners and military aircraft are the most critical element in term of the consequences of bird strike. Aeroengine designers are very keen to have their products pass several FOD tests. These test include high-speed water jets that simulate rain as well as debris, dust, sand, hail, snow, ice, and bird strike.

7.7.1 Certification Regarding Bird Strike

Bird strike can be critical for aircraft engines, especially for its first rotating element, which is the fan in turbofan engines or the compressor in turbojet engines. Federal Aviation Requirements (FAR 33, and its revised version on September 2000) list the necessary engine certification that ensures maximum human safety (either civilian passengers or military personnel) in cases of a bird strike.

The optimum diameter of the air intake for a jet engine depends on the test procedure as well as the size, and weight of the birds shot into the engine by a gas cannon [4].

During ingestion tests employing medium and small birds, several birds are aimed simultaneously at critical regions of the fan blades, such as the blade tips and at the intake of the core engine, as described in the Table 7.1.

The impact of birds can also be simulated by using artificial birds, such as gelatin blocks. All test objects are shot into the running engine by a gas cannon with a speed of up to $500 \, km \, h^{-1}$, which resembles the maximum speed assumed in most international regulations.

This value is much lower than the usual cruise speed of jet-powered aircraft (800–$1000 \, km \, h^{-1}$). However, since most bird strikes happen during takeoff and approach to the airport, then such a test speed ($\leq 500 \, km \, h^{-1}$) is an appropriate choice.

Similarly, European engine certification requirements (CS-E 800) are based on information from various bird strike accidents and are continuously changing based on the perception of the hazard [7]. It is illustrated in Figures 7.18–7.20. A summary for European engine certification requirements (CS-E 800) is given here:

1. Ingestion of a single large flocking bird (1.85–2.50 kg, depending on engine inlet throat area) must not reduce the thrust below 50% of take-off rated value. Also, the engine should be able to maintain this specific thrust for a period greater than 20 minutes (i.e. test schedule) (Figure 7.18).
2. Ingestion of medium flocking birds (whose number and mass depend on the engine throat diameter) must not reduce thrust to less than 75% of take-off rated value (Figure 7.19).
3. Ingestion of small birds (whose number depends on the engine throat diameter and a mass of 0.85 kg each) must not cause a power reduction of less than 25% (Figure 7.20). The mass requirement of 0.85 kg is chosen based on the fact that most small birds hazardous to aviation, such as the starling, have a mass in the range 0.72 kg to 0.83 kg [7].

Table 7.1 Original and revised FAR 33 engine certification bird weight and quantity requirements for engines [4, Chapter 12].

Engine	Inlet area (square inch)	Revised certification standard (September 2000)		Original certification	
		Large-bird quantity and weight	Medium-bird quantity and weight	Large-bird quantity and weight	Medium-bird quantity and weight
JT8	2290	1 of 6.05 lb	1 of 2.53 lb plus 3 each of 1.54 lb	1 of 4.0 lb	4 each of 1.5 lb
RB211	4300–5808		1 of 2.53 lb plus 6 each of 1.54 lb		
CFM56	2922–4072				
PW2037/2043	4902				
JT9D	6940	1 of 8.03 lb	3 each of 2.53 lb		
CF6	6973				
V2500	3217	1 of 6.05 lb	1 of 2.53 lb plus 4 each of 1.54 lb		
PW4000	6940–7854	1 of 8.03 lb	4 each of 2.53 lb		

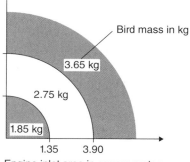

Engine inlet area in square meters

Figure 7.18 Certification requirements for large bird ingestion.

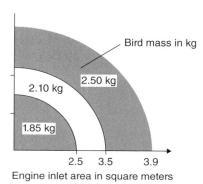

Engine inlet area in square meters

Figure 7.19 Certification requirements for large flocking bird ingestion (applicable only to engines with an engine inlet area greater than 2.5 m^2).

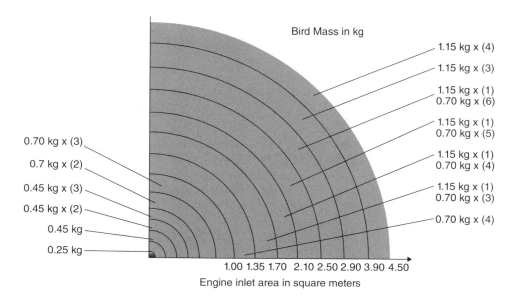

Engine inlet area in square meters

Figure 7.20 Certification requirements for multiple medium-sized flocking birds.

7.7.2 Typical Damage to Turbofan Modules

Figure 7.3 illustrates a typical turbofan engine fastened in a test section and ready for bird strike testing. Dismantling and inspection of the engine after shooting a large bird into its inlet will display one or more of the following findings:

1. damage to some fan blades;
2. damage to several fan exhaust guide vanes;
3. damage to several rotor blades of the first stage of the high-pressure compressor;
4. deformation of the combustor ring;
5. damage to several blades of the low-pressure turbine.

7.8 Helicopters

Birds which weigh less than 4 lb constitute over 90% of all bird strikes with helicopters. The 4 lb bird has been used as a standard for FAA certification of birdproof windshields. No catastrophic failures due to bird strikes have occurred for transparencies designed to this criterion.

Most of the bird strike test facilities for helicopters use a setup similar to that shown in Figure 7.21. The principal features are identical to those described earlier in the testing of fixed-wing aircraft. Its main components are [12]:

- a compressed air source;
- a storage tank;
- a quick-opening valve;
- a breech section for loading;
- a large smooth-bore barrel and timing gates;
- special sabots to fit within the cannon;
- a sabot stripper to remove the sabot before bird impact;
- provision for high-speed photography.

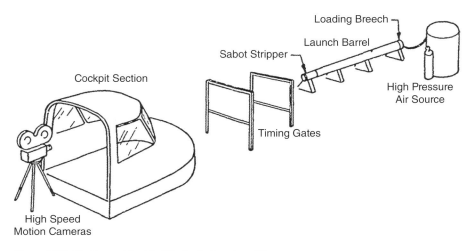

Figure 7.21 Bird impact test facility for helicopters [4].

The birds used are usually freshly killed chickens.

Tests should be conducted on representative, full-size structures to assure that:

- impact shall not cause injury to the flight crew;
- windscreen visibility shall not be reduced more than 50%;
- structural damage shall not cause a mission to abort;
- solid portions of the bird shall not enter the cockpit and cause injury to the crew or prevent them from performing their normal duty.

The Southwest Research Institute (SwRI) conducts safety related full-scale impact testing for helicopters with birds [13]. These tests validate the dynamic simulations of material response to a full-size bird strike.

A large compressed gas gun (LCGG) is employed in bird strike testing. It has the following data:

- pressure vessel rated to 275 psi;
- electronic control to remotely fill and operate;
- 35 ft long, 6 inch diameter one-piece barrel;
- laser alignment;
- 2.2 lb (1 kg) and 4 lb (1.8 kg) birds, 170–310 knots (90–160 m s^{-1}) for ASTM F330-10 testing;
- custom sabot design for high repeatability;
- recessed pit area for high shot lines (~19 ft);
- speeds from 60 knots for rotary-wing FAA Airworthiness Certification up to >550 knots for fixed-wing United States Air Force (USAF) canopy qualification testing.

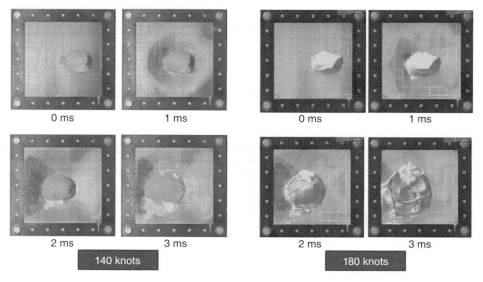

Figure 7.22 An aluminum target inclined 45° as impacted by a 2.2 lb bird at speeds of 140 knots and 180 knots. *Source:* Courtesy of Southwest Research Institute [13].

Three-panel materials are:

- composite (TORAY 48-ply P707AG-15)
- transparent (0.5 inch [12.7 mm] Makrolon PC) and
- aluminum (0.25 in [6.35 mm] 2024-T3).

Figure 7.22 illustrates images taken from high-speed video test data for an aluminum target with 45° slope as impacted by a 2.2 lb bird at two speeds of 140 and 180 knots. The successive figures shown illustrate the deformation of the impacted bird within a very short period of 2 ms [13].

References

1 Read, C.J. (2004). Standard Test Method for Bird Impact Testing of Aerospace Transparent Enclosures. ASTM F330-89. https://www.astm.org/DATABASE.CART/HISTORICAL/F330-89R04.htm (accessed 10 January 2019).

2 Speelman, R.J. III, Kelley, M.E., McCarty, R.E., and Short, J.J. (1998). Aircraft Birdstrikes: Preventing and Tolerating, International Bird Strike Committee, IBSC 24/WP 31 http://www.int-birdstrike.org/Slovakia_Papers/IBSC24%20WP31.pdf (accessed 10 January 2019).

3 Guida, M. (2008). Study, Design and Testing of Structural Configurations for the Bird-Strike Compliance of Aeronautical Components, Ph.D. Thesis, Department of Aerospace Engineering, University of Naples "Federico II". http://www.fedoa.unina.it/3190/1/Guida_doctoral_thesis_Study_Design_and_Testing_of_Structural_Configurations_for_the_bird-strike_compliance_of_aeronautical_components.pdf (accessed 10 January 2019).

4 (2004). *Sharing the Skies: An Aviation Industry Guide to the Management of Wildlife Hazards*, 2e. Transport Canada https://www.ascendxyz.com/wp-content/uploads/regulatory-requirements/Reference%208_Sharing%20the%20Skies.%20An%20Aviation%20Industry%20Guide%20to%20the%20management%20of%20Wildlife%20Hazards.%20Transport%20Canada..pdf (accessed 10 January 2019).

5 Budgey, R. (2000). The development of a substitute artificial bird by the international Bird strike Research Group for use in aircraft component testing. International Bird Strike Committee , IBSC25/WP-IE3. https://www.researchgate.net/publication/278032204_The_development_of_a_substitute_artificial_bird_by_the_international_Bird_strike_Research_Group_for_use_in_aircraft_component_testing_International_Bird_Strike_Committee_ISBC25WP-IE3_Amsterdam (accessed 10 January 2019).

6 Bird Strike Damage & Windshield Bird Strike, ATKINS Final Report, 5078609-rep-03, Version 1.1. https://www.easa.europa.eu/system/files/dfu/Final%20report%20Bird%20Strike%20Study.pdf (accessed 29 December 2018).

7 Maragakis, I. (2009). European Aviation Safety Agency, Safety Analysis and Research Department Executive Directorate, Bird Population Trends and Their Impact on Aviation Safety 1999–2008.

8 European Aviation Safety Agency (2007). Certification Specifications for Large Aeroplanes, CS-25: Amendment 3. https://www.easa.europa.eu/system/files/dfu/

agency-measures-docs-certification-specifications-CS-25-CS-25_Amdt-3_19.09.07_Consolidated-version.pdf (accessed 10 January 2019).

9 Russian Aviation (2015). The bird strike tests of the new Russian MS-21 airliner have been completed. http://www.ruaviation.com/news/2015/8/6/3392/?h (accessed 10 January 2019).

10 Lockheed Martin F-35 Lightning (2016). https://en.wikipedia.org/wiki/Lockheed_Martin_F-35_Lightning_II#/media/File:USS_America%27s_Test_F-35_Flight_Operations.jpg

11 Owens, S.D., Caldwell, E.O., and Woodward, M.R. (2009). Birdstrike Certification Tests of F-35 Canopy and Airframe Structure. http://www.f-16.net/forum/download/file.php?id=13106 (accessed 10 January 2019).

12 Kay, B.F. (1979). *Helicopter Transparent Enclosures, Volume I – Design Handbook.* Stratford, CT: Sikorsky Aircraft Division https://apps.dtic.mil/dtic/tr/fulltext/u2/a065268.pdf (accessed 10 January 2019).

13 Bigger, R., Freitas, C.J., Chocron, S., and Mathis, J. (2016). Experimental Methods in Crash and Impact for Simulation Validation, FAA Workshop on Use of Dynamic Analysis Methods in Aircraft Certification, Blacksburg, VA. https://www.niar.wichita.edu/niarfaa/Portals/0/Experimental%20Methods%20in%20Crash%20And%20Impact%20for%20Simulation%20Validation.pdf (accessed 10 January 2019).

8

Numerical Simulation of Bird Strike

8.1   Introduction

Aircraft bird strikes have been prevalent since the early days of flight in 1903. The number of strikes reported to the Federal Aviation Administration (FAA) annually has increased 5.8-fold from 1851 in 1990 to a record of 10726 in 2012. During this 23-year period (1990–2012), 131096 wildlife strikes were reported to the FAA. Birds were involved in 97% of the reported strikes [1]. On average, every bird strike incident results in a loss of 121.7 flying hours and costs US\$ 32495. Bird strike costs amounted to an economic loss of about US\$ 350 million in 2012, and more than 500000 downtime hours, excluding other monetary losses such as lost revenue, the cost of putting

Bird Strike in Aviation: Statistics, Analysis and Management, First Edition. Ahmed F. El-Sayed.
© 2019 John Wiley & Sons Ltd. Published 2019 by John Wiley & Sons Ltd.

passengers in hotels, re-scheduling aircraft, and flight cancelations. The total economic loss per year is thus estimated to be around US$ 1.28 billion [2].

In collisions between a bird and an aircraft, the point of impact (any forward-facing edge of the vehicle, such as a wing leading edge, nose cone, jet engine cowling, and engine nacelle or inlet) is the most critical point. The same applies for bird strike with the first rotating parts of air-breathing engines, namely the propeller, fan, or compressor. Therefore, the international certification regulations require all forward-facing components to prove a certain level of bird strike resistance before they can be employed in an aircraft. This has been described earlier in this book [3, 4]. Bird strike tests provide a direct method to examine the bird strike resistance. However, experimental techniques are very costly (a single bird shooting test can cost several thousands of dollars) and also do not provide enough information about the structural response [5]. Thus, to shorten the design cycle and reduce costs, numerical simulations are often used to examine and assess a structure's response to bird strike.

Some early studies used force–impulse models and semi-empirical equations to predict the average force during a bird strike event [6]. Though these equations are based on momentum conservation, they may not account for the complex structural interaction that occurs between the bird and the target, and thus cannot be used to determine the imposed target damage. LS-DYNA [7] is an appropriate software that can simulate the bird–structure interaction.

Numerical techniques provide the designer with a wide range of useful data, such as stress distribution, displacements, and 3D visual observations of structure deformation. In addition, numerical techniques enable a parametric study of different materials, geometry, and impact speed (magnitude and direction). The process is easy and costs only computational time. With the use of numerical simulations, the number of required experimental tests is decreased significantly [8].

The impact of a bird on an aircraft is an impact dynamics problem with a very short duration. Compared to quasi-static simulations, bird strike modeling includes several complexities such as transient intense loads, fluid–solid interactions, strain rate effects, large deformations, and severe element distortion [9].

Since the duration of the event is very short (and, therefore, the materials are highly accelerated), the effect of material inertia cannot be ignored. Stress waves are so strong that their transmission and mutual interactions must be taken into account.

Due to the different natures of solids and fluids, proper modeling of the force interactions and impulse transfers poses a considerable problem. For a large deformation, the impacted structure will be in the plastic phase rather than the elastic phase. Moreover, shear deformation and rotational inertia must be considered in the plate or beam theories.

Finite element (FE) modeling of bird strikes dates to the late 1970s, when a few research studies were carried out to examine the response of the canopies and windshields of airplanes impacted by birds.

In 1977, the linear FE code IMPACT was not able to yield acceptable results for bird strikes. In 1978, the computer program MAGNA (Materially and Geometrically Nonlinear Analysis) was developed and two years later was applied to the impact of birds on the transparent components of aircraft [10]. In the mid-1980s, the bird-proof capability of several aircraft parts, including laminated composites [11], horizontal stabilizers [12], and turbine engine fan blades [13], were also investigated by means of finite elements.

In the 1970s and 1980s, simulations were first performed of bird–canopy impacts to determine the mesh refinement necessary to match with the available experimental results for deformations due to bird strike. It was found that only a very fine mesh proved satisfactory. However, this mesh could not be used in the computer simulations of the 1980s due to limited computer core-storage and its high cost. Therefore, the amount of numerical/computer work undertaken on bird strike in the 1980s was relatively little and remained so until around the year 2000, when the computing power of computers became sufficient for the accurate simulation of complex problems with relative ease [14].

Researchers found that a bird could strike an aircraft component with its head, tail, rear, or wings, and any of these orientations might produce a different effect on the response of the component. Four substitute bird models were introduced and the best substitute model was chosen to capture the pressure and force exerted by a real bird when impacting from different orientations [15].

To properly model the structural deformation and material failure of both the bird (as the projectile) and the impacted structure, the implemented FE package must be able to handle the following topics [5]:

- strain rate effects;
- excessive element distortion;
- accurate prediction of load distribution in the interface between the bird and the target;
- modeling the fasteners (usually accomplished by defining the single node-to-node tie);
- reliable material behavior at very severe conditions;
- geometrical and material nonlinearity;
- fast and accurate solutions;
- energy dissipation.

8.2 Numerical Steps

The finite element method (FEM) is based on the idea of discretizing a complex geometry into a large number of small pieces with simple basic geometries. These simple geometries are usually triangles and quadrilaterals for 2D analyses and hexahedral and tetrahedral for 3D analyses. The vertices of the constructing pieces (or elements) are known as the nodes of the FE model.

Analyzing a physical problem in a FE packages is done in three main steps [16].

- Pre-processing: the FE model which is going to be solved is prepared.
- Solution: the prepared FE model is solved. Additional controlling parameters are usually used in this step to avoid solution instabilities.
- Post-processing: the quantities of interest are obtained from the solved problem. These include stress, strain, displacements, velocities, accelerations, resultant forces, visualization, etc.

The solution step is usually completely handled by the FE package and, therefore, the user does not have to deal with it.

In some FE packages, such as ANSYS and ABAQUS, all the three steps are contained in a single graphical user interface (GUI). In others packages the user must use different programs for each step. For example, in LS-DYNA [17], which is the most prevalent program in bird strike analysis, the solution program is separate from the pre- and the post-processing consoles.

In some cases, the user may prefer using other more recent programs for the better handling of a specific step. As an example, the program HyperMesh is preferred by many users due to its high efficiency in discretization of very complex geometries.

8.2.1 Pre-processing

Pre-processing is the process in which all the data required by the solver program is prepared. In a pre-processing procedure, the following items must be done:

- Creating the 2D or 3D computer-aided design (CAD) model of the physical object. In very complex geometries, the geometrical model can be prepared in a professional CAD program, and then exported into the FE pre-processing program.
- Specifying the material models that determine the behavior of each material type under different loading conditions.
- Specifying the material constants of each material.
- Selection of the suitable element formulation for different parts of the problem.
- Discretization of the model with well-shaped elements.
- Applying the loads and boundary conditions to the corresponding nodes.

8.2.2 Solution

In the solution step, first, the local stiffness matrix and the force vector of each element is constructed. By combining all the local stiffness and force matrices, a global stiffness matrix and a global force vector are constructed. The program then finds the displacements at the nodes by solving the set of equations:

$$\{F\} = [K]\{\delta\} \tag{8.1}$$

Generally, the FE solvers solve a field problem by using either explicit or implicit techniques. If the load is high enough that the material yielded or the displacements are larger than static displacements, the FE problem cannot be solved using Eq. (8.1) in a single step. In these conditions, the total duration of time through which the material is going to be deformed must be divided into very small time increments (or time steps).

In an explicit analysis, the response of the material in each time increment is calculated. At the end of each time increment, the geometry and material behavior of the structure is updated and the stiffness matrix elements are modified for the next increment. In the next time increment, the updated stiffness matrix is used for obtaining the new displacements [16].

Similarly, for implicit FE, the solver uses the updated stiffness matrix in each time increment. The difference between the two analysis methods is that the implicit solver uses Newton–Raphson iterations to ensure that the internal forces of the structure are in equilibrium with the externally applied loads. In the explicit method, the new state is calculated based on the state of the system in the previous time step, while in the implicit method the new state of the system is calculated by considering the state of the system at both the previous and the next time steps. Therefore, the implicit solutions usually give better accuracy.

However, implicit solutions are usually very time-consuming and therefore not suitable for problems with very large deformations. Implicit analyses are mainly useful for the transient problems which take a long time to occur. On the other hand, explicit analyses are useful for high-velocity problems with very short durations in which the effect of shock waves is important [17].

Smaller time steps lead to greater accuracy, whilst larger time steps yield faster solutions. To obtain sufficiently accurate results, the time step must be as small as the time it takes for a shock wave to travel through the smallest element in the model. Therefore, the higher the velocities of the objects in a simulation are, the smaller the required time step will be. Time steps which are small enough ensure that all the energies related to the propagating shock waves are considered in the analysis [17].

In a bird strike problem with a bird initial velocity of about $100\,\mathrm{m\,s^{-1}}$, and a bird constructed with 5000–15 000 elements, then time step will be of the order of 0.1–1 µs. Even if the bird strike lasts for 1–10 ms, the small size of the resulting time steps means that the solver needs to solve the problem incrementally more than a thousand times. Moreover, since the implicit analyses try to enforce the equilibrium of the system by Newton–Raphson iterations, the high nonlinearity in the system makes the convergence in the implicit analysis more difficult. Therefore, using explicit analysis in a bird strike problem is well established.

8.2.3 Post-processing

In the post-processing step, the user can obtain the required data from the results of the solver program after the completion of the solution. The most common results that are important for a bird strike problem [16] are:

- pressure distribution in the interface surface;
- the permanent plastic deformations in the impacted component;
- the eroded elements of the structure;
- the time history visualizations.

8.3 Bird Impact Modeling

Bird impact on aircraft is a soft body impact. It requires the density of a fluid, its viscosity, the shape of the bird (the projectile), and the length-to-diameter ratio to be precisely selected.

First, we will discuss its geometry and material and then impact dynamics will be considered.

8.3.1 Modeling the Geometry and Material of Birds

Extensive studies have been performed for bird models using the following geometries (Figure 8.1):

- straight-ended cylinder [18];

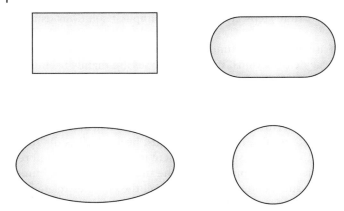

Figure 8.1 Various shapes for birds.

- hemispherical-ended cylinder [19];
- ellipsoid [20];
- sphere [8].

The hemispherical-ended cylinder is the most common geometry, whilst the straight-ended cylinder is the least common one. A study focused on the first three configurations mentioned above at various length-to-diameter aspect ratios [21] showed that altering the aspect ratio has little effect on the pressure profile or impulse diagrams.

The four geometries were modeled in another numerical study [8]; they impacted rigid targets from their axial and lateral sides. The length-to-diameter ratio was set to two for all the geometries except the sphere. The results of straight and lateral impacts were compared between the four geometries as well as the experimental tests [22] to find the most appropriate bird-substitute geometry. In a straight-ended cylinder, a larger area of the target plate is affected by very high pressure peaks and, consequently, the straight-ended cylinder is the most damaging bird substitute in an impact. In some research work, the bird is simulated by a solid cylinder with truncated coned ends [23].

However, real birds have body shapes completely different from the simple geometries discussed above. The inner part of a bird's body is also very complex, consisting of organs with different shapes and materials having various strengths. Examples are the bones, muscles, and internal cavities, containing different organs such as the lungs, as well as the beak. Such materials have differing strengths. Also, each bird species has different strengths and geometrical properties.

However, there are a few works in which the researchers have implemented more complex bird geometries which more closely resemble real bird shapes. A bird model with geometry like the bufflehead duck was examined by Hedayati and Ziaei-Rad [8]. A real bird may strike an aircraft component with its head, tail, bottom, or wings. Any of these orientations might produce a different effect on the response of the aircraft component. Since all birds have different body shapes and sizes, four substitute bird models were introduced. It was concluded that the impact from the bottom side of a bird is the most damaging scenario, while the tail-side impact is the least dangerous one. It was also found that for the tail-side impact scenario, a hemispherical ended

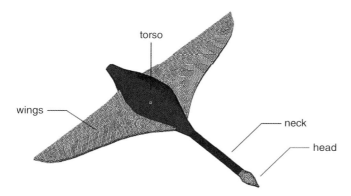

torso

wings

neck

head

Figure 8.2 Schematic of the multi-material bird model. *Source:* reproduced from [6] with permission.

cylinder shows the best results, while for the bottom-side impact scenario, an ellipsoid can be the best candidate for the bird-substitute model.

Figure 8.2 illustrates a complex multi-material bird model representing a Canadian goose weighing 3.6 kg [6].

The results of this multi-material bird model were compared with the results of a hemispherical-ended cylinder. It was observed that for the multi-material bird model with its neck extended, the target was pre-stressed due to the impact of the bird's head and neck prior to the torso. Based on the results [6], the authors asserted that modeling bird organs, e.g. head, neck, wing, etc. in addition to the bird torso is necessary when modeling large birds such as a Canadian goose. They stated that consideration of other body parts in bird models can have significant effects on damage initiation of aircraft components, which in turn can determine the final failure situation.

Finally, the bird material needs to be considered. A bird behaves like a fluid in a high velocity impact. The density values used in multi-material model are as listed in Table 8.1.

8.3.2 Impact Modeling

Bird impact behavior on rigid targets can be divided into four main stages [17], as illustrated in Figure 8.3.

1) Initial shock stage, when the first compression wave is formed and propagates back into the bird material.
2) Pressure decay (release) stage, when the bird's particles located in its periphery tend to be released radially.
3) Steady-state stage, when bird particles flow in fixed streamlines in space.
4) Pressure termination stage, when all the bird particles have reached the target surface and the pressure descends to zero.

Table 8.1 Densities of the sections of the multi-material bird model [6].

Element	head	neck	torso	wings
Density (kg/m^3)	900	1500	1150	845

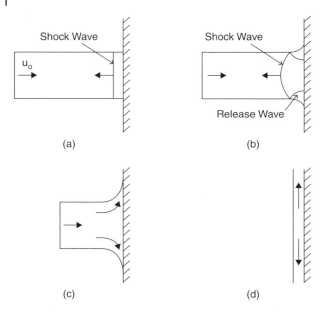

Figure 8.3 Impact phases of soft body. *Source:* reproduced from [17] with permission.

In the first stage, the bird impacts the surface with an initial velocity, u_o. Bird material near the surface is instantaneously brought to rest, which creates a shock wave (called the Hugoniot shock). This shock wave is parallel to the surface of impact and propagates down the bird's body. The initial pressure peak at the impact point is called the Hugoniot pressure, P_H. The shock wave results in a significant pressure difference on the surface. This results in material flow on the un-impacted free surface.

As the material accelerates outwards, a release wave is formed (stage 2), which propagates toward the center of the impact, interacting with the shock wave. The interaction decreases the pressure at the impact point and represents the second stage of impact. After several cycles of the release waves interacting with shock waves, the material begins to flow steadily (stage 3). The pressure continues to decrease as the release waves interact with the shock waves. Eventually (stage 4) steady flow is established at which point the material flows entirely radially. In this regime, a steady stagnation point is reached, which has a maximum at the center of impact.

In the first phase of impact theory (Figure 8.3a), the generated high pressure for a subsonic velocity is given by the water hammer equation:

$$P = \rho_O c_0 U_o \tag{8.2}$$

where ρ_O is the material initial density, c_0 is the speed of sound in the material, and U_o is the velocity of the bird material.

As the shock propagates (Figure 8.3b), the impact velocity increases beyond a subsonic range and a modified version of water hammer equation is used to obtain the Hugoniot pressure (P_H) as follows:

$$P_H = \rho_O U_s U_o \tag{8.3}$$

where U_S is the velocity of the shock in bird material.

Equations (8.2) and (8.3) are appropriate for perfectly rigid materials. Since aircraft and engine materials are of the compliant type which may behave either elastically or plastically depending on impact load, a modified version is suggested by Wilbeck [25] and expressed by Eq. (8.4), where subscripts (P and T) stand for projectile (here the bird) and target (here the impacted surface).

$$P_c = \rho_P U_{SP} U_o \left[\frac{\rho_T U_{ST}}{\rho_P U_{SP} + \rho_T U_{ST}} \right] \tag{8.4}$$

In Eq. (8.4) P_c is the pressure at the impact zone.

Moreover, the initial peak pressure depends only on densities and velocities and not on the length or cross-sectional area of the projectile. However, as reported by Wilbeck [25], bird-substitute materials, such as a gelatin and water mixture, have higher peak pressure than a real bird because of its high shock velocity and lack of porosity compared to the anatomical structure of birds (which includes bones, lungs, and other cavities that reduce the bird density).

After the initial phase of the impact, there is the decay pressure. The shock pressure loading decreases with time and with distance from the center of the impact region; the shock wave emerging at the lateral free surfaces of the body is followed by a set of release waves. The radial pressure distribution is given by Eq. (8.5):

$$P_r = P_c e^{-\frac{Kr}{R(t)}} \tag{8.5}$$

where P_c is obtained from Eq. (8.4), K is a constant, r is the radial distance from the center of the impact region, and $R(t)$, a function of time, is the maximum contact radius at time t.

Finally, the steady (stagnation) pressure on the impacted surface is given by:

$$P_s = K \rho_0 U_o^2 \tag{8.6}$$

which is independent of bird shape. This steady-state pressure is 10–30% of the peak Hugoniot shock pressure at the center of the impacted area [25], and the factor $K = 0.5$ for incompressible fluid and for compressible it approaches unity.

The bird is assumed to be a fluid body; thus, the bird impact is assumed to be a hydrodynamic impact. Birds do not bounce. Impact begins when the leading edge of the bird first touches the target. The impact continues until the trailing edge of the bird reaches the target and there is no further bird material flowing onto the target. If the bird does not decelerate during impact, then this squash-up time, T_s is given by:

$$T_s = \frac{L}{U_0} \tag{8.7}$$

where L is the length of the bird. Since the duration time of bird strike is very short, the effect of strain rate should be taken into consideration. Figure 8.4 illustrates the typical variation of impact pressure with time for any case of a high-speed bird impact on a rigid plate.

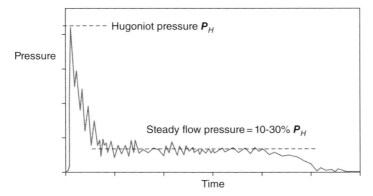

Figure 8.4 Variation of impact pressure versus time. *Source:* reproduced from [17] with permission.

8.4 Numerical Approaches for Bird Strike

In present bird strike studies, the bird and impacted structure are first modeled. The compatibility conditions are maintained in the contact surface. FE representation of the aircraft structure of various modules is rather straightforward since their mechanical properties and geometries are known. On the contrary, the bird has complex properties and these are largely unknown. The appropriate numerical method is that which leads to a solution close to the experimental solution.

Currently, there are four modeling methods used for the impact damage analyses: Lagrangian, Eulerian, arbitrary Lagrangian Eulerian (ALE), and smooth particle hydrodynamics (SPH).

8.4.1 Mathematical Models

The governing equations are conservation of mass, conservation of momentum, and constitutive relationship of the material, and these are essential to solve the soft body impact problem. These three major equations are solved by LS-DYNA to obtain the velocity, density, and pressure of the fluid particle for a specific position and time.

Conservation of mass:

$$\frac{D\rho}{Dt} + \rho \nabla . \bar{V} = 0 \tag{8.8}$$

where ρ is the density, and $\bar{V}$ is the velocity vector.

Conservation of momentum:

$$\nabla \sigma + \rho \mathbf{b} = \frac{D}{Dt}\left(\rho \bar{V}\right) \tag{8.9a}$$

where σ is the stress tensor, **b** is the body force vector per unit mass, and ρ is the density. For the impact problem, no body forces are considered: thus, **b** = 0. Further, it is assumed that for equilibrium the normal compressive stress acting in the contact interface balances the normal pressure exerted for the body over the target within the contact area

[26]. Also, it is assumed that the shear stresses are neglected. Thus, the conservation of momentum becomes:

$$-\nabla P = \frac{D}{Dt}\left(\rho \bar{V}\right)$$ (8.9b)

State equation:

$$P = P(\rho)$$ (8.10)

8.4.2 The Lagrangian Method

It was the first, and the most common, approach to model bird strike. The Lagrangian method is the default approach for discretizing solid parts in FE packages. The structure is divided into many discrete finite elements forming the finite element mesh. The mesh follows the material, i.e. one material per element during the entire simulation. In other words, this mesh is associated with particles in the material under examination. Therefore, each node of the mesh follows an individual particle in motion.

This model can follow the distortions of the bird material, and to some extent, the break-up of bird material into debris [27].

Figure 8.5 shows the grid in the Lagrangian method. The initial grid is shown in Figure 8.5a while the distorted grid after time Δt is displayed in Figure 8.5b. Figure 8.6 illustrates the Lagrangian bird deformation after time t_1.

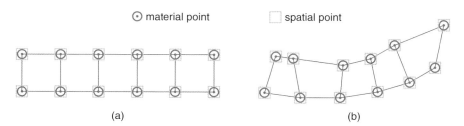

Figure 8.5 Lagrangian description.

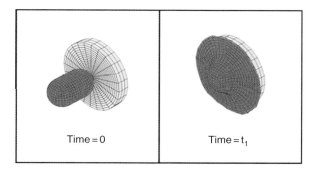

Figure 8.6 Lagrangian bird deformation. *Source:* reproduced from [24] with permission.

Table 8.2 Advantages and disadvantages of the Lagrangian formulation in bird strike applications.

Advantages	Disadvantages
Simple imposition of boundary conditions as the boundary nodes remain on the material interface	May result in loss of bird mass due to the fluid behavior of the bird, which causes large distortions in the bird
Easy ability to track the time-history of each particle since spatial points and bird material points are coincident	Excessive element distortion leading to instability. It can cause severe mesh tangling which in turn results in notorious, non-physical negative volumes due to elements folding over themselves
Low computational time and thus low cost	Severe element distortion specially in explicit solvers may increase the number of time steps

Advantages and disadvantages of the Lagrangian method are given in Table 8.2.

To overcome the solution instabilities caused by small time steps, the highly distorted elements are eliminated. However, if the number of distorted elements are relatively large compared to the total number of bird elements, both of the conservation of mass and the conservation of energy principles are violated, leading to unrealistic results [28].

8.4.3 The Eulerian Approach

The Eulerian approach is an alternative numerical modeling approach for overcoming the difficulties related to Lagrangian bird modeling. Since at high pressures, the bird body behaves like a fluid, the Eulerian approach, which is prevalent in modeling fluid dynamic problems, would be helpful. The computational grid is fixed in space and the material passes through the mesh, as illustrated in Figure 8.7. Thus, a fixed void mesh is created in space and some of the cells are filled by bird material at the points where the bird must be present. As the bird material travels into the space, some cells become hollow and some others become filled with the bird material. Here also there will be no mesh distortion.

The Eulerian approach requires a relatively fine mesh to yield sufficiently accurate results. This significantly increases the computational time required.

The advantages and disadvantages of the Eulerian approach for modeling bird strike are identified in Table 8.3.

Figure 8.8 illustrates the Eulerian bird deformation after time t_1.

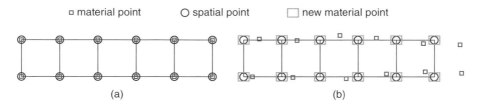

(a)   (b)

Figure 8.7 Eulerian description.

Table 8.3 Advantages and disadvantages of the Eulerian formulation in bird strike applications.

Advantages	Disadvantages
Complex material can be modeled with no limit to the amount of deformation or distortion	Diffusion in the material boundaries using sophisticated interface tracking implementations
Large time steps	Tracking of the material history is difficult as the solid stress and strain tensors must be transported from cell to cell
Can easily handle multiple materials in one cell	A wide-meshed domain is necessary, which increases computational time

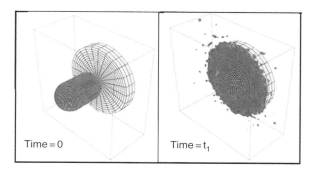

Figure 8.8 Lagrangian bird deformation. *Source:* reproduced from [24] with permission.

8.4.4 The Arbitrary Lagrangian Eulerian (ALE)

The ALE method is similar to the Eulerian method in modeling the bird strike. In the ALE method, the material is free to move with respect to the mesh (Figure 8.9). This method combines the best features of purely Lagrangian and Eulerian approaches by circumventing their inherent disadvantages [52]. The ALE method benefits from the stability of the Eulerian method and the low computational cost of the Lagrangian method.

The only difference between ALE and Eulerian methods is that in the ALE method the mesh is not necessarily fixed in space and can move in the direction of the projectile's center of gravity. Therefore, the large number of void space elements in the Eulerian method can be reduced in the ALE method, and the void elements can be limited to

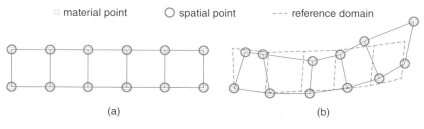

Figure 8.9 ALE description.

Table 8.4 Advantages and disadvantages of the ALE method in bird strike applications.

Advantages	Disadvantages
Smaller grid structure size compared to the Eulerian method	Inaccurate results in high deformations
No excessive material distortion for the bird material compared to the Lagrangian method	High CPU time as high numbers of cells are required
Better accuracy for contact modeling compared to the Eulerian method	The user must be experienced to set the mesh motion that best suits the problem to minimize the mesh distortions and so obtain the best results

only the regions around the bird model [29]. The ALE coupling algorithm is relatively difficult to utilize (since it is based on the penalty method), and the user must be familiar with a large set of controlling parameters to achieve acceptable results. For its advantages and disadvantages refer to Table 8.4.

8.4.5 Smoothed Particles Hydrodynamics (SPH)

The SPH method is a mesh-less or grid-less Lagrangian technique for modeling transient fluid motion using a pseudo-particle interpolation method. It is effective and accurate in predicting the deformation of fluids. Since the SPH method is meshless, no additional elements are required to be defined to represent the void space, in contrast with the Eulerian or ALE methods.

In the SPH method, the fluid is represented by a cloud of moving small particles, each one being an interpolation point, where all the fluid characteristics are known. An interpolation function, namely a kernel function, is used to find the desired quantities for all the particles. The kernel function is active only over a limited zone for each interpolation point. Each interpolation node is given a mass, and the values of state variables for the node are determined based on the mass of the node itself and the masses and distances of the adjacent nodes in its zone.

In the SPH analysis it is important to know which particle will interact with its neighbors as the interpolation depends on these interactions. Therefore, a neighboring search technique has been developed, as shown in Figure 8.10 [30]. The influence of a particle

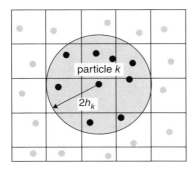

Figure 8.10 Active domain around a particle k in the SPH method.

(k) is established inside of a sphere of radius $2h_k$ where h is the smoothing length. In the neighboring search, it is also important to list, for each time step, the particles that are inside that sphere.

The advantages and disadvantages of the SPH method in bird strike applications are summarized in Table 8.5.

The particles in the SPH method are equally spaced in both parallel and normal planes to the impacted target. Four shapes for projectiles resembling birds are illustrated in Figure 8.11.

8.5 Case Study

In the following sections the previous numerical techniques will be applied for different configurations; namely the leading edge of wing/tails, the windshield, fan blades, and the helicopter rotor.

Table 8.5 Advantages and disadvantages of the SPH method in bird strike applications.

Advantages	Disadvantages
Does not require a numerical grid; thus, problems with irregular geometry can be solved	Some issues in the areas of stability, consistency, and conservation
Efficient tracking of material deformations and history-dependent behavior	High computational demands, both in memory and in CPU time
Does not suffer from the normal problems of grid tangling and element distortion in large deformation problems	High computational cost
High accuracy	

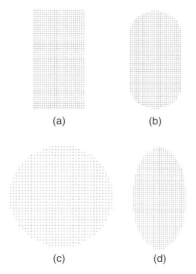

(a) (b)

(c) (d)

Figure 8.11 Cross sectional geometry for SPH numerical models of four projectile shapes: (a) right circular cylinder, (b) hemispherical cylinder, (c) sphere, (d) ellipsoid.

8.5.1 Leading Edges of Wing/Tail

Impact of birds with the leading edge configurations, which may be of a wing or horizontal/vertical tail, has been thoroughly examined [17]. Two numerical methods were adopted; namely, Lagrangian and SPH impact simulations. For the Lagrangian method the bird was idealized as a cylindrical bullet of homogenous material.

In the SPH method, the bird model geometry was approximated as a right circular cylinder with hemispherical end caps. The bird was modeled as a projectile having a length to diameter ratio of two. Its weight and density were 3.86 kg and 950 kg/m^3 respectively.

8.5.1.1 Wing

Figure 8.12 shows the finite element model of the leading edge structure, the truss, and the Lagrangian bird model (here as a right cylinder).

The impact velocity was 129 m s^{-1} and no structural penetration was present. Based on the numerical results, the stress on the inner skin had a peak value equal to or greater than three times its allowable stress.

When the Lagrangian approach was employed, the numerical model of the bird geometry (shape and dimensions) and structural material (constitutive law and equation of state, EOS) was performed using explicit FE codes (MSC/Dytran code). However, for the SPH approach the LS/Dyna code was employed and the bird was modeled using the hemispherical end caps.

Figure 8.13 illustrates the initial shapes and successive deformations of both the bird and the structure in four impact steps using the Lagrangian method (0, 1, 2, and 3.6 ms).

Figure 8.14 illustrates the initial shapes and successive deformations of both the bird and the structure in four impact steps using the SPH method (0, 1, 2, and 3.6 ms).

Figure 8.15 illustrates another simulation for bird strike on a composite leading edge using the SPH method. Bird impact led to a penetration of the skin after only 2 ms. On impact the bird material starts to spread but most of the mass enters the wing interior

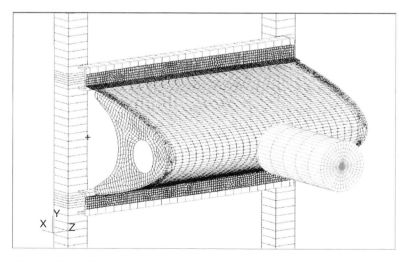

Figure 8.12 Leading edge bay FE model. *Source:* reproduced from [17] with permission.

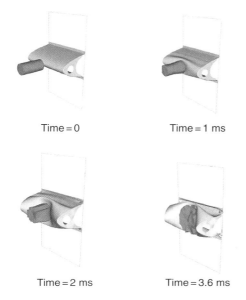

Figure 8.13 Four impact steps using Lagrangian method (0, 1, 2, and 3.6 ms). *Source:* reproduced from [17] with permission.

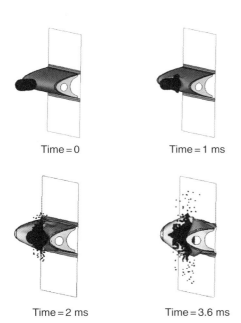

Figure 8.14 Four impact steps using SPH method (0, 1, 2, and 3.6 ms). *Source:* reproduced from [17] with permission.

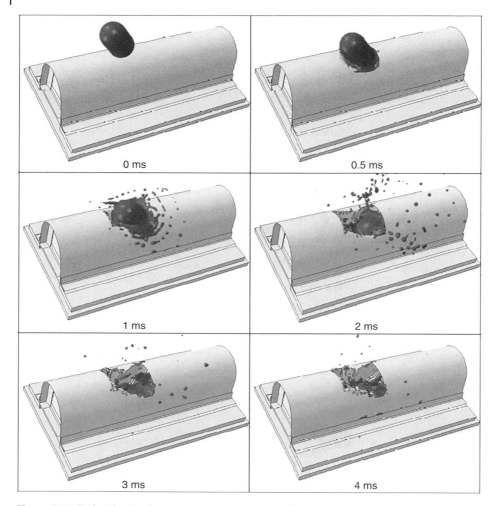

Figure 8.15 Bird strike simulation on composite wing leading edge. *Source:* reproduced from [24] with permission.

and continues to damage the structure. This could lead to penetration of fuel tanks and other sensitive components of the aircraft.

Similar analysis of a composite wing leading edge subjected to impact has been studied by Ericsson [31]. Comparisons with experimental data showed that delamination can be modeled using contact between shell layers in FE analysis.

The SPH model was also used in an impact on an aircraft wing leading edge structure by McCarthy et al. [32]. The SPH bird model was able to capture the breakup of the bird into debris particles after its collision with the wing leading edge structure, something that was difficult to accomplish using the Lagrangian method. Figure 8.16 simulates the bird impact when the leading-edge slat is deployed. Penetration of the structure is seen.

A detailed simulation for bird strike on an aircraft wing was performed by Walvekar [33]. A bird simulated by the SPH model impacted the leading edge of an aircraft wing

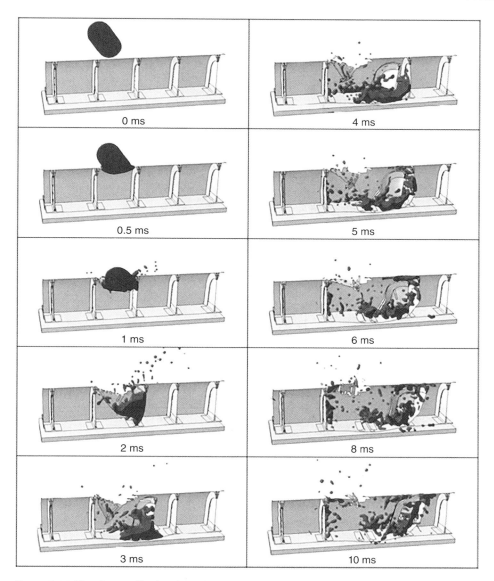

Figure 8.16 Time lapse of bird strike with deployed leading edge slats collision event starting in the top left corner. *Source:* reproduced from [24] with permission.

(with the wing having different nose radius values of 12.7, 62, and 64 mm) at a constant angle of 30° and impact speed varying from 146 to 427 knots.

A 4 lb bird impacted the leading edge of a wing having a nose radius of 64 mm at a velocity of 341 knots. The bird strike at different instants is shown in Figure 8.17.

Figure 8.18 illustrates the leading edge damage at the end of simulation where physical damage to the leading edge skin, leading edge rib, spar, and diaphragm are observed.

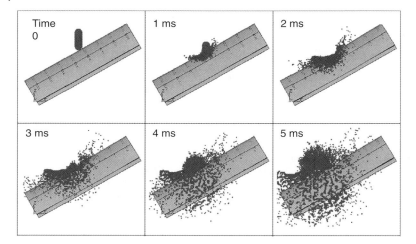

Figure 8.17 Frames at different instants of time at impact velocity of 341 knots. *Source:* reproduced from [33] with permission.

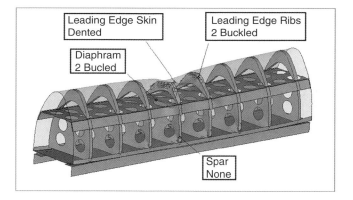

Figure 8.18 Damage of leading edge having 64 mm nose radius as impacted at a velocity of 341 knots. *Source:* reproduced from [33] with permission.

Bird impact behavior on the wing leading edge of a military aircraft was investigated by Lammen [34]. The bird was modeled using the SPH method. The contact forces for impacts at speeds of 100 and $160\,\mathrm{m\,s^{-1}}$ are shown in Figure 8.19.

8.5.1.2 Vertical Tail

A study looking at bird strike on the leading edge of fin of the vertical tail of the C27J cargo aircraft was investigated by Guida [17]. Lagrangian finite element analysis was used to calculate the stress distribution on the fin. Figure 8.20 shows that the damage produced to the structure is local, and that only small portion of the leading edge is damaged by the strike while the rest of structure remains in the elastic range. The fringe plot shows the stress distribution in MPa. Results outlined that a peak stress of 400 MPa is located in the contact zone. The simulations show that the leading edge is able to withstand the specified impact without the bird penetrating in the structure.

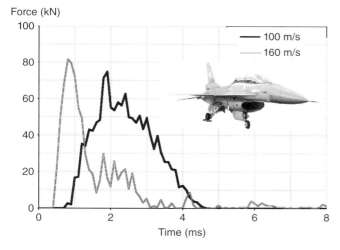

Figure 8.19 Impact force of the bird on the leading edge of a military aircraft [34]. *Source:* Courtesy of NLR.

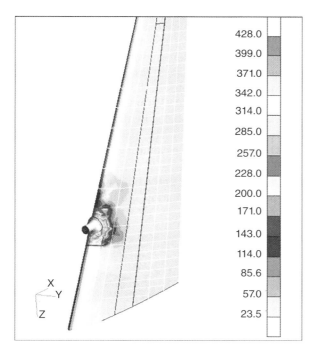

Figure 8.20 Max stress on the C27J fin using Lagrangian Approach. *Source:* reproduced from [17] with permission.

8.5.1.3 Horizontal Tail

There are several innovative designs for a bird strike-resistant composite leading edge for the horizontal tail of a transport aircraft based on a novel application of composite materials with high energy-absorbing characteristics: the tensor-skin concept was examined by Ubels et al. [35]. Three improved leading edge structures with different energy-absorbing tensor concepts were manufactured. Bird strike tests with a 4 lb synthetic bird at impact velocities around 100 m s^{-1} were performed. Finite element models were developed to examine several innovative designs for a bird strike-resistant, composite leading edge.

Figure 8.21 shows typical simulation results for one of three structures; namely, NLR-LE-2 impacted normally midway between the ribs by a 1.82 kg synthetic bird at 100 m s^{-1} impact velocity. This case represents a bird kinetic energy of 9.1 kJ at a typical landing speed of a commuter aircraft when bird impacts are likely and where the leading edge is required to absorb bird energy to prevent damage to the wing spar.

There is extensive damage and fracture in the cover and carry plies with unfolding of the tensor strips. The SPH bird model flows in a realistic way over the leading edge structure, with part of the bird trapped in the tensor ply. (The tensor ply contain folded loops between the ribs, which unfold when a relatively high lateral load is applied to the skin, e.g. in the case of bird strike.) Simulation results compare well with the damaged structure from the experimental test.

8.5.2 Sidewall Structure of an Aircraft Nose

The coupled SPH and finite element (FE) method is used to predict the dynamic responses of the sidewall structure of an aircraft nose [15]. The numerical simulation of the bird strike test is carried out using the explicit finite element software PAM-CRASH. Figure 8.22 shows the finite element model, which consists of three parts: the bird, the sidewall structure, and the clamping fixture. The geometry of the bird is idealized as a right cylinder having a diameter of 106 mm and a length of 212 mm to keep the standard length to diameter ratio of 2 : 1. The idealized bird is modeled by the SPH method and the Murnaghan EOS. The sidewall structure consists of ribs, pad plate, skin-1, and skin-2 and these parts are modeled by shell elements. The clamping fixture is modeled by solid elements.

The SPH simulation captures the deformation behavior of the bird, which was broken into debris and streamed across the structure. The deformation and fracture process of

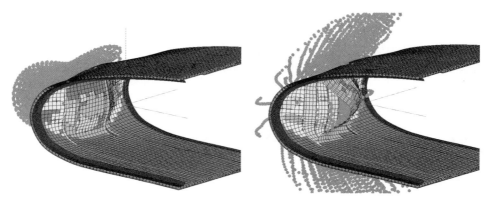

Figure 8.21 Bird strike simulation of NLR-LE-2 with 1.82 kg bird at 100 m s^{-1} [35]. *Source:* Courtesy of NLR.

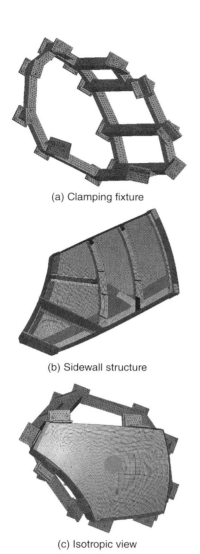

(a) Clamping fixture

(b) Sidewall structure

(c) Isotropic view

Figure 8.22 Finite element models of structure and bird [15].

the sidewall structure is accurately predicted by the numerical simulation, as illustrated by Figure 8.23. The process of the bird strike test is captured by a high-speed camera recording, which lasts about 3 ms. A remarkable agreement between the simulation results and the experimental results recorded by high-speed camera is noticed.

8.5.3 Windshield

Based on several statistics, as described in Chapter 2, bird strike with the windshield represents some 13% of the total bird strike accidents/incidents for civil aircraft and 20% for military aircraft. Consequently, extensive numerical simulation of bird strikes with the windshield is performed [9, 36–40].

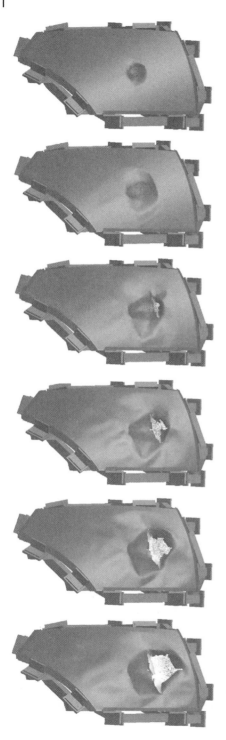

Figure 8.23 Simulated bird strike process [15].

The nonlinear finite element program LS-DYNA is used to numerically simulate bird strike on an aircraft windshield [36]. A parametric study for such analysis was established to investigate the responses of the windshield under three different bird velocities at three sites [36]. The bird is represented by a cylinder with a hemisphere at each end and the contact-impact coupling algorithm is used in this study. The windshield is represented by a Lagrangian reference configuration, whereas the bird is represented by a multi-material ALE formulation to avoid too large a deformation. The geometry and mesh of the bird and the full FE model are shown in Figure 8.24. A bird of mass 1.8 kg was modeled by material model 3 of DYNA (PLASTIC_ KINEMATIC). The bird is set to strike on three locations (A, B, and C in Figure 8.24) of the windshield surface with three different initial velocities of 125, 150, and 160 m s^{-1}.

Numerical simulations proved that when the bird strikes on the central point, the windshield will be broken at 160 m s^{-1} (Figure 8.25), while there is no damage at this position at 150 m s^{-1}.

Another detailed numerical simulation for bird impacts on windshields was performed by developing a nonlinear FE model in commercially available explicit FE solver AUTODYN [41]. An elastic-plastic material model coupled with maximum principal strain failure criterion was implemented to model the impact response of the windshield. The numerical model was validated with published experimental results and further employed to investigate the influence of various parameters on the dynamic behavior of the windshield. The parameters include the mass, shape, and velocity of the bird, the angle of impact, and the impact location (Figure 8.26). Based on numerical results, the critical bird velocity and failure locations on windshield were also

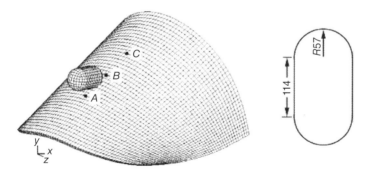

Figure 8.24 The FE model of a bird strike and three impact locations [36].

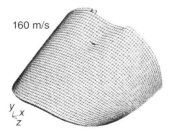

Figure 8.25 Broken windshield when the bird strikes at point C [36].

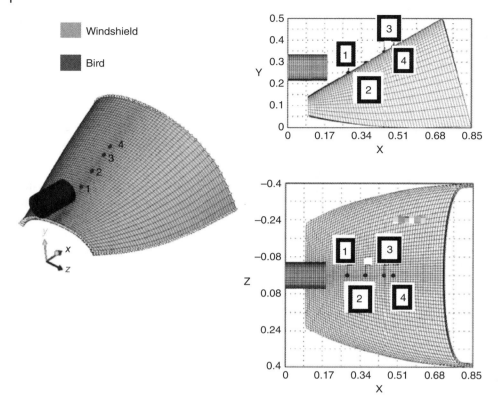

Figure 8.26 Finite element model of bird impact on windshield: isometric view (left), side and top views (right) [41].

determined. The results show that these parameters have a strong influence on the impact response of the windshield. Bird velocity and impact angle were amongst the most critical factors to be considered in windshield design.

Figure 8.27 shows the effect of impact velocity on the deformation of the windshield

8.5.4 Fan

An aircraft engine is a very critical module from point of view of bird strike. Most current engines, in both civil and military applications, are turbofan engines [42, 43]. The fan is the first rotating module of the turbofan engine. Birds may seriously damage one or more blades of the fan, resulting in loss of power or successive damage of the other

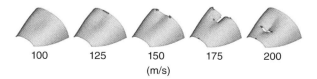

Figure 8.27 Deformation modes of windshield at different impact velocities [41].

engine modules. Moreover, as described in Chapter 2, bird strike with engines represents 40–50% of bird strikes with different aircraft components. Engines must confirm the certification requirements listed by the FAA and/or the European Aviation Safety Agency (EASA). As described in Chapter 7 and this chapter, engine certification tests are difficult and costly. These expenses can be significantly reduced by using a numerical simulation of the fan blade/bird strike problem in the design of a jet engine. However, only rather limited number of papers have been found that deal with bird strike analysis of rotating engine fan blades [44–47].

The common technique for such simulations is to model the bird as a solid ellipsoid with material properties similar to water. Numerical simulation of fan blade bird strike analysis using LS-DYNA Lagrangian, SPH, and ALE approaches to model the bird are adopted by Ryabov et al. [44]. The main objectives of the investigations are to compare the results obtained by means of different approaches and to find out the advantages and disadvantages of every approach.

A computer model of the fan blades and three computer models of the bird were created for the numerical simulations. A view of the blades and the bird models prepared for the ALE approach is represented in Figure 8.28.

The Lagrangian model of the bird consists of 90 000 solid finite elements, the ALE mesh consists of 720 000 finite elements, and the bird is initially defined in ~25 000 ALE cells.

The total CPU time for calculation using the code (Lagr3) was 5 hours. Figure 8.29 shows sequential views of the interaction of the bird and blades. Analysis of the results

Figure 8.28 Computer model (ALE approach). *Source:* reproduced from [44] with permission.

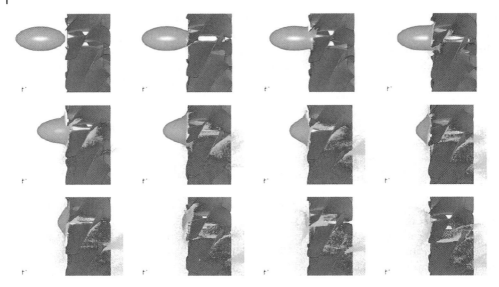

Figure 8.29 Sequential views of the interaction of the bird and blades (Lagr3). *Source:* reproduced from [44] with permission.

shows that there is no penetration of the blades into the bird's material and all the eroding nodes stay active in the contact.

A detailed study for bird strike of aircraft engine fans was prepared by Shultz and Peters [45]. The study used shell elements for the blades, rigid solid elements for the hub, shell elements for the shroud, and solid ALE elements for the Euler mesh to transport the bird material.

The structural mesh was constructed as described here. The blades' airfoil meshes were created in ANSYS with Shell 163 elements (Belytschko-Tsay formulation). The thickness at each node is defined appropriately for a smooth mesh transition. The hub mesh was created as rigid Solid 164 elements. The hub mesh shares nodes with the blade mesh at the root of the airfoils. The shroud mesh was created with Shell 163 elements (Belytschko-Tsay formulation). The shroud mesh density was similar to the mesh density of the blade at the tip. The bird material was simulated with an Euler mesh. The Euler mesh is a stationary mesh which transports the bird material within it. At the beginning of the analysis the initial bird material elements had a fraction equal to 100% and the rest of the elements was 0%. As the analysis progresses, the bird material translated through the mesh, changing the volume fractions of the elements which contained it.

Thus, when a shell element (Lagrangian mesh) from the airfoils swept through a brick element (Euler mesh) of the air that had a volume fraction greater than a threshold value, an additional set of calculations were triggered which was the mechanism of momentum transfer between the bird material and the airfoils.

Two bird was launched and after 0.059398 s, the first bird hit the fan, as illustrated in Figure 8.30. Figure 8.31 illustrates the instant when the second bird impacted the fan blade after a time = 0.085094 s.

Another study treating two birds impact with a seventeen-blade fan were considered in [46]. In view of the results obtained, the SPH bird model is the most feasible model for the analysis.

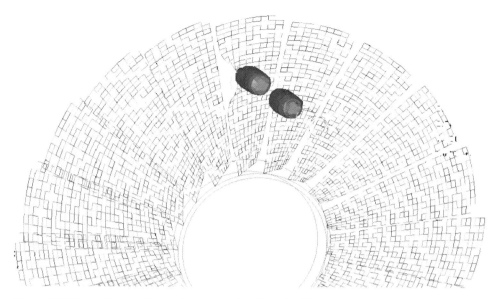

Figure 8.30 Blade damage due to the impact of the launched first bird after a time of 0.059398 s. *Source:* reproduced from [45] by permission.

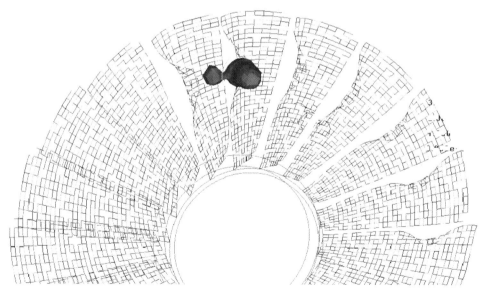

Figure 8.31 Blade damage due to the impact of the second bird after a time of 0.085094 s. *Source:* reproduced from [45] by permission.

A detailed study for bird strike with the fan was carried out by Selezneva [47]. The focus of the study was to implement the SPH method and investigate the different aspect ratios and EOS of the bird. The study had the following data: bird mass = 0.7 kg, bird speed = 75 m s^{-1}, fan rotational speed =10 000 rpm. The fan is of the intermediate size and the study focused on the blades only. Moreover, only 12 blades were considered

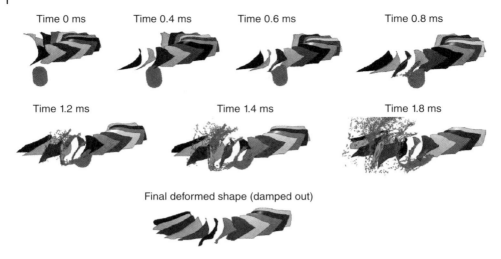

Time 0 ms   Time 0.4 ms   Time 0.6 ms   Time 0.8 ms

Time 1.2 ms   Time 1.4 ms   Time 1.8 ms

Final deformed shape (damped out)

Figure 8.32 Bird/blade interaction at successive time steps. *Source:* reproduced from [47] by permission.

in the study. Fully integrated shell elements were also employed in this study. The impact model took into consideration the stresses arising from blade rotation. The bird was modeled by a cylinder with two hemispherical ends. Moreover, the number of particles used in SPH varied from 10 000 to 110 000 to assess the sensitivity of model. The sensitivity study showed that the minimum acceptable number of particle was 36 000.

Bird/blade interactions at successive time steps are illustrated in Figure 8.32.

8.5.5 Helicopter Windshield

Helicopters are more at risk than other aircraft because of their low altitudes and comparatively very large windows. Birds may penetrate the windshield of a helicopter, the person who is struck can be seriously injured, and the bird may fall into the cockpit. Therefore it is particularly important that the design and construction of a helicopter cockpit safeguards against bird strike. As with fixed-wing aircraft, this requires precise information on expected loading, which can only be delivered by a combination of very extensive experiments and simulations. There is considerable interest, therefore, in minimizing experimental time and effort and in using more numerical models. This is only possible if the predictive quality of the simulation is high enough to rate the results as trustworthy and reliable.

Several research works used the mesh-free or grid-less SPH method for the solution of the Euler equation. These studies made it possible to model the failure patterns of comparatively brittle material with good precision [48].

Numerical simulations for the windshields in the light helicopters category is discussed by Cwiklak and Grzesik [49]. Both classic Lagrangian-based modeling and the SPH approach were adopted. A bird of mass 1.8 kg was modeled in three different shapes; spherical ($D = 153.53$ mm), cylindrical (having its length as twice the diameter with $D = 106.45$ mm), and cylindrical with hemispherical ends ($L = 2D = 226.24$ mm).

The Agusta A-109 helicopter manufactured by Agusta Westland was selected. Its glass windshield has a thickness of 3.81 mm. The windshield was modeled by two methods. These are (i) the shell elements Belytschko-Tsay with five degrees of freedom in

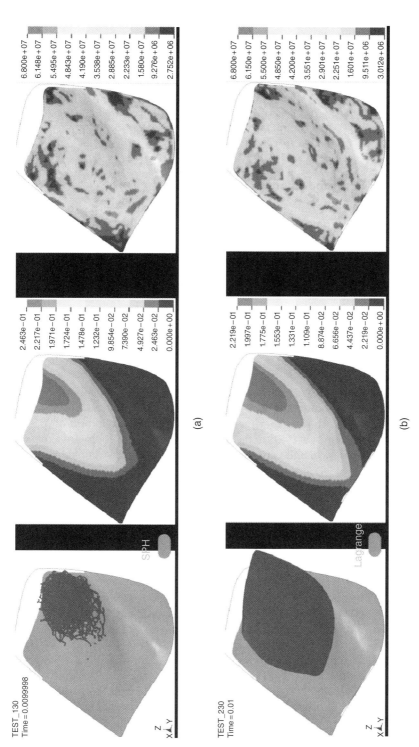

Figure 8.33 Windshield deformation, displacement, and stresses after 0.01 s [49]: (a) SPH method, (b) Lagrange method. *Source:* reproduced from [49] with permission.

each node and (ii) a solid element generator LS-PrePost, which formed a solid element mesh by adding thickness to the existing shell elements.

Numerical analysis exploited the computational code LS-DYNA v.970. The bird impact velocity was $285\,\mathrm{km\,h^{-1}}$ ($79.167\,\mathrm{m\,s^{-1}}$) while the total analysis time was adopted as 10 ms, which is divided into several time steps.

Figure 8.33 illustrates the deformations of the solid elements of the windshield caused by a bird model strike for the last time step. The middle column illustrates the total displacement contours (in m) while the right column illustrates the reduced stress contours (in Pa).

8.5.6 Helicopter Rotor and Spinner

Few published work handle numerical treatments for bird strike with either the rotor or spinner of helicopters. Famous helicopter manufacturer AgustaWestland (AW) Italy used the code RADIOSS and SPH bird formulation to simulate a case of bird strike onto a rotor blade [50]. The blade is made of composite material and it has been modeled with solid and shell elements to simulate the fillers and the skins of birds. To simulate the clamp, a layer of rubber wrapping the airfoil has been modeled, using the /MAT/ OGDEN card for the material.

Another numerical analysis for bird strike with the spinner of a helicopter rotor was performed by AgustaWestland [51]. The spinner of AgustaWestland AW609 (formerly the Bell/Agusta BA609) was treated. The ALE technique in LS-DYNA was employed to predict the composite failures and structural impact performance for the spinner and rotor assembly.

The bird was simulated by a cylindrical shape having two hemispherical ends surrounded by air, which is modeled as 1-point ALE solid elements. These numerical results were compared with the experimental results for shooting a bird of mass 1.8 kg using a compressed gas gun at 240 knots ($127.3\,\mathrm{m\,s^{-1}}$) impact velocity into the spinner control unit. Both numerical and experimental results matched to a great extent.

References

1 Katukam, R. (2014). Comprehensive Bird Strike Simulation Approach for Aircraft Structure Certification, CYIENT White paper. http://www.cyient.com/fileadmin/cyient. com/media/downloads/white-papers/141128-WP-Aero-Bird-Strike-Sim-EN.pdf (accessed 11 January 2019).

2 Allan, J.R. and Orosz, A.P. (2001). The Costs of Birdstrikes to Commercial Aviation, Paper 2, page 2. 2001 Bird Strike Committee-USA/Canada, Third Joint Annual Meeting, Calgary, AB. http://digitalcommons.unl.edu/birdstrike2001/2/ (accessed 11 January 2011).

3 Heimbs, S. (2011). Computational methods for bird strike simulations: a review. *Computers and Structures* 89 (23–24): 2093–2112.

4 Mithun, N. and Mahesh, G.S. (2012). Finite element modelling for bird strike analysis and review of existing numerical methods. *International Journal of Engineering* 8 (1): 1–8.

5 Guida, M., Marulo, F., Meo, M., and Russo, S. (2013). Certification by birdstrike analysis on C27J fullscale ribless composite leading edge. *International Journal of Impact Engineering* 54: 105–113.

6 McCallum, S.C. and Constantinou, C. (2005). The influence of bird-shape on bird-strike analysis, 5th European LS-DYNA users conference. https://www.dynalook.com/european-conf-2005/ALE_FSI_SPH_I/Mccallum.pdf (accessed 11 January 2019).

7 Hallquist, J.O. (2006). *LS-DYNA – Theory Manual*. Livermore, CA: Livermore Software Technology Corporation http://www.lstc.com/pdf/ls-dyna_theory_manual_2006.pdf (accessed 11 January 2019).

8 Hedayati, R. and Ziaei-Rad, S. (2013). A new bird model and the effect of bird geometry in impacts from various orientations. *Aerospace Science and Technology* 28: 9–20.

9 Wang, F.S. and Yue, Z.F. (2010). Numerical simulation of damage and failure in aircraft windshield structure against bird strike. *Materials and Design* 31 (2): 687–695.

10 McCarty, R.E. (1980). Finite element analysis of F-16 aircraft canopy dynamic response to bird impact loading. 21st Structures, Structural Dynamics, and Materials Conference, Seattle, WA, USA. https://doi.org/10.2514/6.1980-804 (accessed 11 January 2019).

11 Sun, C.T. (1972). An analytical method for evaluation of impact damage energy of laminated composites. In: *Composite Materials: Testing and Design (Fourth Conference)*, vol. 617, 427–440. ASTM STP.

12 McCarty, R.E. (1983). MAGNA computer simulation of bird impact on the TF-15 aircraft canopy. In: *Proceedings of the 14th Conference on Aerospace Transparent Materials and Enclosures*, 974–1008. Scottsdale, AZ.

13 Storace, A.F., Nimmer, R.P., and Ravenhall, R. (1984). Analytical and experimental investigation of bird impact on fan and compressor blading. *Journal of Aircraft* 21 (7): 520–527.

14 Blair, A. (2008). Aeroengine Fan Blade Design Accounting for Bird Strike. B.Sc. Thesis, University of Toronto. https://www.mie.utoronto.ca/mie/undergrad/thesis-catalog/237.pdf (accessed 11 January 2019).

15 Jun, L., Yulong, L., Gao, X., and Yu, X. (2014). A numerical model for bird strike on sidewall structure of an aircraft nose. *Chinese Journal of Aeronautics* 27 (3): 542–549.

16 Hedayati, R. and Sadighi, M. (2016). *Bird Strike: An Experimental, Theoretical and Numerical Investigation.* Woodhead Publishing in Mechanical Engineering, Elsevier.

17 Guida, M. (2008). Study, Design and Testing of Structural Configurations for the Bird-Strike Compliance of Aeronautical Components, Ph.D. Thesis, Department of Aerospace Engineering, University of Naples "Federico II". http://www.fedoa.unina.it/3190/1/Guida_doctoral_thesis_Study_Design_and_Testing_of_Structural_Configurations_for_the_bird-strike_compliance_of_aeronautical_components.pdf (accessed 10 January 2019).

18 Salehi, H., Ziaei-Rad, S., and Vaziri-Zanjani, M.A. (2010). Bird impact effects on different types of aircraft bubble windows using numerical and experimental methods. *International Journal of Crashworthiness* 15 (1): 93–106.

19 Langrand, B., Bayart, A.S., Chauveau, Y., and Deletombe, E. (2002). Assessment of multiphysics FE methods for bird strike modelling – application to a metallic riveted airframe. *International Journal of Crashworthiness* 7 (4): 415–428.

20 Budgey, R. (2000). The development of a substitute artificial bird by the international bird strike research group for use in aircraft component testing. International Bird

Strike Committee, ISBC25/WP-IE3, Amsterdam. https://www.researchgate.net/
publication/278032204_The_development_of_a_substitute_artificial_bird_by_the_
international_Bird_strike_Research_Group_for_use_in_aircraft_component_testing_
International_Bird_Strike_Committee_ISBC25WP-IE3_Amsterdam (accessed 11
January 2019).

21 Ivančević, D. and Smojver, I. (2011). Hybrid approach in bird strike damage prediction
on aeronautical composite structures. *Composite Structures* 94 (15–23): 2011.

22 Wilbeck, J. (1978). *Impact Behavior of Low Strength Projectiles*. Wright-Patterson Air
Force Base, Ohio: Air Force Wright Aeronautical Labs https://apps.dtic.mil/dtic/tr/
fulltext/u2/a060423.pdf (accessed 11 January 2019).

23 Shmotin, Y.N. et al. (2009). Bird Strike Analysis of Aircraft Engine Fan. 7th European
LS-DYNA Users Conference. https://www.dynamore.de/en/downloads/papers/09-
conference/papers/H-I-03.pdf (accessed 11 January 2019).

24 Heimbs, S. (2011). Bird Strike Simulations on Composite Aircraft Structures. 2011
SIMULIA Customer Conference https://www.researchgate.net/publication/260022977_
Bird_Strike_Simulations_on_Composite_Aircraft_Structures (accessed 11
January 2019).

25 Wilbeck, J.S. (1977). Impact Behavior of Low Strength Projectiles.Air Force Wright
Aeronautical Labs. Wright-Patterson Air Force Base, Ohio. https://apps.dtic.mil/dtic/
tr/fulltext/u2/a060423.pdf (accessed 11 January 2019).

26 Goyal, V.K., Huertas, C.A., and Vasko, T.J. (2013). Bird-strike modeling based on the
Lagrangian formulation using LS-DYNA. *American Transactions on Engineering &
Applied Sciences* 2: 57–81.

27 Stoll, F. and Brockman, R.A. (1997). Finite element simulation of high speed soft-body
impacts. In: *Proceedings of the 1997 38th AIAA/ASME/ASCE/AHS/ASC Structure,
Structural Dynamics, and Materials Conference*, 334–344. Kissimmee, FL, USA.

28 Nizampatnam, L.S. (1999). Models and Methods for Bird Strike Load Predictions. PhD
Thesis, Aerospace Engineering, Wichita State University. http://soar.wichita.edu:8080/
bitstream/handle/10057/1494/d07030.pdf?sequence=1;Models (accessed 11
January 2019).

29 Jenq, S.T., Hsiao, F.B., Lin, I.C. et al. (2007). Simulation of a rigid plate hit by a
cylindrical hemi-spherical tip-ended soft impactor. *Computational Materials Science*
39 (3): 518–526.

30 Grimaldi, A. (2011). SPH High Velocity Impact Analysis: A Birdstrike Windshield
Application. Doctoral dissertation, Università degli Studi di Napoli Federico II. http://
www.fedoa.unina.it/8221/ (accessed 11 January 2019).

31 Ericsson, M. (2012). Simulating Bird Strike on Aircraft Composite Wing Leading Edge,
M.Sc Thesis, KTH Royal Institute of Technology, Stockholm, Sweden. https://www.
diva-portal.org/smash/get/diva2:561636/FULLTEXT01.pdf (accessed 11 January 2019).

32 McCarthy, M.A., Xiao, J.R., McCarthy, C.T. et al. (2004). Modeling of bird strike on an
aircraft wing leading edge made from fibre metal laminates – part 2: modeling of
impact with SPH bird model. *Applied Composite Materials* 11: 317–340.

33 Walvekar, V., Thorbole, C.K., Bhonge, P., and Lankarani, H.M. (2010). Birdstrike
Analysis on Leading Edge of an Aircraft Wing Using a Smooth Particle Hydrodynamics
Bird Model. In: *ASME International Mechanical Engineering Congress and Exposition*,
vol. 1, 77–87. *Advances in Aerospace Technology* https://doi.org/10.1115/
IMECE2010-37667 (accessed 11 January 2019).

34 Lammen, W.F., van Houten, M.H., and ten Dam, A.A. (2009). Predictive simulation of impact phenomena in aircraft design. Two case studies with common mathematical approach, Report no. NLR-TP-2008-442. https://core.ac.uk/download/pdf/53034125. pdf (accessed 11 January 2019).

35 Ubels, L.C., Johnson, A.F., Gallard, J.P., and Sunaric, M. (2003). Design and testing of a composite bird strike resistant leading edge. Report no. NLR-TP-2003-054. https:// core.ac.uk/download/pdf/53034430.pdf (accessed 11 January 2019).

36 Xinjun, W., Zhenzhou, F., Fusheng, W., and Zhufeng, Y. (2007). Dynamic response analysis of bird strike on aircraft windshield based on damage-modified nonlinear viscoelastic constitutive relation. *Chinese Journal of Aeronautics* 20 (6): 511–517.

37 Boroughs, R.R. (1998). High Speed Bird Impact Analysis of the Learjet 45 Windshield Using DYNA3D. https://doi.org/10.2514/6.1998-1705 (accessed 11 January 2019).

38 Wang, A.J., Qiao, X., and Li, L. (1998). Finite element method numerical simulation of bird striking multilayer windshield. *Acta Aeronautica et Astronautica Sinica* 19 (4): 446–450. [in Chinese].

39 Bai, J.Z. (2003). Inverse issue study of bird-impact to aircraft windshield based on neural network method. PhD thesis, Northwestern Polytechnical University [in Chinese].

40 Wang, F.S., Yue, Z.F., and Yan, W.Z. (2011). Factors study influencing on numerical simulation of aircraft windshield against bird strike. *Shock and Vibration* 18 (3): 407–424.

41 Dar, U.A., Zhang, W., and Xu, Y. (2013). FE analysis of dynamic response of aircraft windshield against bird impact. *International Journal of Aerospace Engineering* 2013: https://doi.org/10.1155/2013/171768.

42 Elsayed, A.F. (2017). *Aircraft Propulsion and Gas Turbine Engines*, 2e. Boca Raton, FL: Taylor & Francis, CRC Press.

43 Elsayed, A.F. (2016). *Fundamentals of Aircraft and Rocket Propulsion*. London: Springer.

44 Ryabov, A.A., et al. (2007). Fan Blade Bird Strike Analysis Using Lagrangian, SPH and ALE Approaches, 6th European LS-DYNA Users' Conference, 29–30 May 2007. https://www.dynalook.com/european-conf-2007/fan-blade-bird-strike-analysis-using-lagrangian.pdf (accessed 11 January 2019).

45 Shultz, C. and Peters, J. (2002). Bird Strike Simulation Using ANSYS LS/DYNA. 2002 ANSYS users conference, Pittsburgh, PA.

46 Castelletti, L.-M.L. and Anghileri, M. (2005). Bird strike influence of the bird modeling on the slicing forces. 31st European Rotorcraft Forum.

47 Selezneva, M. et al. (2012). Modeling of bird impact on a rotating fan: The influence of bird parameters. *FEA Information Engineering Journal* 1 (4): 37–44.

48 Tho, C.-H. and Smith, M.R. (2008). Accurate Bird Strike Simulation Methodology for BA609 Tiltrotor. American Helicopter Society 64th Annual Forum, Montréal, Canada, April 29 – May 1, 2008.

49 Cwiklak, J. and Grzesik, N. (2016). Numerical analysis of bird strike events with the helicopter windshield. *Journal of Kones Powertrain and Transport* 23 (3).

50 Bianchi, F. (2009). Analysis with RADIOSS of a bird strike onto a helicopter blade and onto a rotor control chain. http://altairatc.com/europe/Presentations_2009/Session_11/ AGUSTA_ALtair%20Users'%20conference-2-4_11_09.pdf (accessed 11 January 2019).

51 Tho, C.-H. and Smith, M.R. (2006). Bird strike simulation for BA609 spinner and rotor controls. Proceedings of the 9th International LS-DYNA Users Conference. https://

www.dynalook.com/international-conf-2006/43SimulationTechnology.pdf/view (accessed 11 January 2019).

52 Smojver, I. and Ivancevic, D. (2010). Bird impact at aircraft structure – damage analysis using coupled euler lagrangian approach. *IOP Conference Series: Materials Science and Engineering* 10 (1): http://iopscience.iop.org/article/10.1088/1757-899X/10/1/012050 (accessed 11 January 2019).

9

Bird Identification

9.1   Introduction

Reporting wildlife strikes is crucial to the continuing effort of bird strike prevention. Bird/wildlife strike statistics cannot be properly interpreted unless the species struck is known [1]. Identification of bird species involved in bird/aircraft strikes is an important part of the overall assessment and management of wildlife mitigation at airports. Knowing the exact species means that the size, behavior, and ecology of the bird in question is known and is key to tracking species trends as well as focusing preventative measures. Species identification assists engineers in designing critical modules such as windscreens and engines to be more resilient to bird strike events [2].

In biological classification, taxonomic rank is the relative level of a group of organisms and is order as species, genus, family, order, class, phylum, kingdom, and domain, where species is the lowest rank and domain is the highest rank [14]. Species and genus are written in italic type. The scientific system of naming "kinds" of birds revolves around the species level. For example, *Turdus migratorius* (American robin) contains genus and then the species names [15].

(In the Biological Species Concept, BSC, the definition of "species" is based on species being reproductively isolated from each other. Thus, distinctive geographical forms of the same "kind" of bird are usually combined as one species. However, the

Bird Strike in Aviation: Statistics, Analysis and Management, First Edition. Ahmed F. El-Sayed.
© 2019 John Wiley & Sons Ltd. Published 2019 by John Wiley & Sons Ltd.

Phylogenetic Species Concept, PSC, says that different geographic forms of the same basic "kind" of bird should be treated as distinct species since these forms have evolved separately and have unique evolutionary histories. Obviously, the PSC is less restrictive than the BSC. There would be many more species of birds under the PSC than under the BSC.)

Bird evidence (bird remains) is typically collected by field personnel (e.g. pilots, maintenance crew, airport biologists) according to procedures established by each agency [16]. Figure 9.1 illustrates feather remains on engine blades, while Figure 9.2 illustrates remains on a helicopter. Bird/wildlife remains recovered following strikes are often fragmentary, but even the smallest feather fragments can be identified. Blood smears can be used to identify species using different a range of methods, including DNA analysis.

Airports should ensure that all bird remains are identified as completely as possible, given the facilities at their disposal. Feathers can reveal the birds' species, age, and sex.

The Smithsonian Institution (SI) Feather Identification Lab (Washington, DC) is a highly specialized laboratory that has processed several thousand cases of bird species identification annually, from either whole or fragmentary feather material, since the

Figure 9.1 Collection of remains of bird strike from an aircraft engine [2].

Figure 9.2 Collection of remains of bird strike from a helicopter [2].

1960s [3]. Scientists in the Smithsonian Institution can identify birds using one or more of the following methods:

a. visual methods from feather characters: size, shape, color, and texture;
b. microscopic methods, looking at node shape, barbule length, distribution, and pigmentation patterns [4–7];
c. DNA barcoding methods [8].

Identification can be confirmed by circumstantial evidence from the bird strike event, such as its date, location, and eye witness reports. These help to validate the final identification.

9.2 Collecting Bird Strike Material

Following a collision between a bird and an aircraft, small feather fragments and blood stains are usually the only evidence that a bird was involved in the incident.

9.2.1 Feathers

If the whole bird is available, pluck a variety of feathers (breast, back, wing, tail). If only a partial bird is found, collect a variety of feathers that are colored or patterned. If only feathers are available, send all available material.
The instructions that need to followed [3] are:

- do not cut feathers from the bird, as the downy part at the base of the feathers is needed;
- do not use any sticky substance, so do not use tape or glue.

9.2.2 Tissue/Blood ("Snarge")

The word "snarge" means the residue smeared on an airplane after a bird/plane collision.

9.2.2.1 Dry Material
With dry material, use one of the following methods:

- scrape or wipe off the material into a clean re-closable bag;
- wipe the area with a pre-packaged alcohol wipe;
- spray the area with alcohol to loosen material then wipe with clean cloth/gauze.

Note: do not use water, bleach, or other cleansers as they destroy DNA.

9.2.2.2 Fresh Material
Use the appropriate method from the following:

- wipe the area with an alcohol wipe and clean cloth/gauze;
- apply fresh tissue/blood to an FTA® DNA collecting card;

The instructions for the collecting process are:

1. always include any feather material that is available;

2. include a copy of the report (AFSAS, WESS, or FAA 5200-7);
3. always secure all remains in a re-sealable plastic bag.

9.3 Reporting and Shipping

When reporting and shipping bird strike remains, there are several guidelines for collecting and submitting feather or other bird/wildlife remains for species identification [9]. The guidelines listed below will help to maintain species identification accuracy, reduce turn-around time, and maintain a comprehensive database.

1. Collect and submit remains as soon as possible.
2. Provide a complete report information regarding the incident:
    a. fill out form FAA 5200-7 (a copy can be printed from https://wildlife.faa.gov/strikenew.aspx);
    b. mail the report with the material;
    c. provide contact information if the species identification is needed.
3. Collect as much material as possible (Figure 9.3):
    a. pluck/pick a variety of feathers from the wings, tail, and body;
    b. include any feathers with distinct colors or patterns;
    c. include any downy "fluff;"
    d. include beaks, feet, and talons if possible;
    e. if only a small amount of material is available then all of it should be sent;
    f. do not send the whole birds;

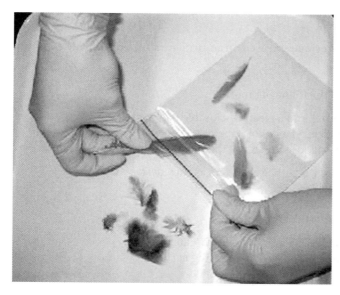

Figure 9.3 Collecting a variety of whole feather material [2].

g. do not cut off feathers, as cutting will remove the downy region needed to aid in identification;

h. do not use any sticky substances, such as tape or sticky notes to attach feathers.

4. Place the material in clean plastic/zip-lock bags.

5. Mail the report and material to the Smithsonian Institution. Their report will be sent to the FAA, Office of Airport Safety and Standards.

There are two options for shipment of bird strike remains – the regular US postal service and overnight/priority shipping [2]. If the bird strike has caused damage then priority or overnight shipping is recommended. It is important to securely package the material and use the correct address. Shipments can be labeled "safety investigation material."

If the material is sent from outside the US, the Feather ID Laboratory should be first contacted for special treatment and shipping requirements.

Feather identification is free of charge to all US aircraft owners/operators at all US airport operators, and to any foreign air carrier if the strike happened at a US airport. The turnaround time for species identification is usually 24 hours from receipt. The reports and species identification information are then sent to the USDA, NWRC, in Sandusky, Ohio.

9.4 Methods Used to Identify Bird Strike Remains

The method used to identify bird strike remains depends on what kind of material is available.

9.4.1 Examination by Eye

Eye examination of bird is a macroscopic method. The results depend on the skill of the examiner and the condition of the feather remains. Eye examination was the employed method in Europe before 1978. On average, 26 samples, mostly large remains, were sent annually to the Zoological Museum Amsterdam, of which 80% could be recognized [10].

If there is a whole bird or a partial carcass, identification can be based on the physical characteristics traditionally used when viewing birds in the wild – including size, color, and pattern. Wings, feathers, feet, and beaks can then be compared with the bird specimens of the museum collections to make a final identification. This approach is also used when samples only include loose or fragmented feathers.

Experienced ornithologists examine feathers by eye to determine the species or group involved; the findings can be verified through comparison with specimens in a museum collection. It is estimated that 75% of struck birds can be identified using this technique.

Figure 9.4 illustrates the procedure when identifying the feather remains of a bird strike. The worker is matching an unknown tail feather with a museum specimen of a killdeer (*Charadrius vociferus*).

The Smithsonian Feather Identification Lab identifies birds involved in wildlife strikes by comparing their remains with the museum specimens. Figure 9.5 illustrates feather specimens from a short-eared owl and a peregrine falcon.

Figure 9.4 Matching an unknown tail feather with a museum specimen of a killdeer [2].

9.4.2 Microscopic Examination

Identification of samples consisting of small feather fragments, blood, and tissue involves examination under a microscope [2]. This technique can be used to identify the family or genus of bird involved, but usually does not provide a species identification.

When preparing for microscopic analysis, the downy barbs are cut off close to the shaft of the feather and are sandwiched between the microscope slide and a coverslip. These are then glued together along the edges. If the feather remains are very dirty or greasy, they are agitated in a container of warm water to which a liquid soap or detergent has been added. After being washed, the feathers are rinsed and then dipped in alcohol for a few seconds to speed drying. Dirty or twisted feathers can usually be restored to their original shape by this procedure.

The microscopic features of the downy part of a feather are unique for different groups of birds (e.g. duck, raptor, or passerine). Looking at this fluffy area of the feather can provide valuable clues to narrowing down the species identification. Figure 9.6 illustrates the microscopic structure of a meadowlark feather, which is a type of many passerines.

Scanning electron microscopy is an advanced microscopic method that can be used to detect differences and identify at a more precise taxonomic level. It can be used on samples of the Anatidae and Charadriidae bird families that could not be identified by other means [10]. Feathers are examined using a scanning electron microscope. Magnification ranges from 500 to 1000 times, which permits a detailed examination of

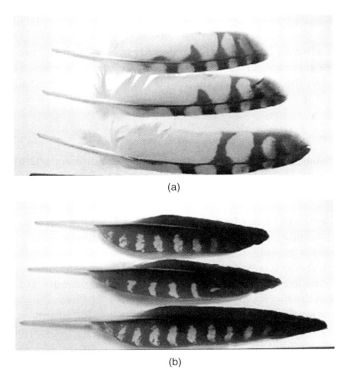

(a)

(b)

Figure 9.5 Feather specimens in the Smithsonian Feather Identification Lab [2]: (a) short-eared owl, (b) peregrine falcon.

Figure 9.6 The microscopic structure of meadowlark feather [2].

the barbs, barbules, nodes, and internodes. Precise measurements of the various components of the feathers are recorded for comparative purposes and analysis. Within each family, only insignificant differences were found between the various species, and these differences are not sufficient to allow identification at the genus level [10].

Bloodstains can be used in microscopic identification [12]. Identification of avian blood/tissue from aircraft is of paramount importance. However, as the samples are usually very small and are often heavily contaminated and poorly preserved, usual simple laboratory procedures are full of discrepancies. Using other methods, such as chemical examination of blood by the benzidine/orthotoluidine test under simulated air crash situations, is also useful [12].

9.4.3 Keratin Electrophoresis

Identification of feathers by visual means leaves some unidentifiable samples, particularly at the lower taxonomic levels. Optical microscopy and scanning electron microscopy can improve results, but still about 25% of the samples cannot be identified below the family level.

Electrophoresis of proteins extracted from feather keratin, used previously in taxonomic research, can provide reliability and repeatability in identifying feather remnants from any source. Keratin proteins provide a fingerprint which is consistent within a particular species. Protein extraction has been refined and standardized, as has the methodology for electrophoresing feather protein concentrates. The sample must be at least 10 mg for identifications to the species level [11].

There is little individual variation and differences between species are significant and can often be assessed visually. When gels are scanned with a laser densitometer, the differences between each keratin profile are more obvious and can be measured. The values of the keratin profile curve can be used for separating closely related species. Their results [11] indicate a high success rate and precision in identification, exceeding the results obtained by other means for samples that cannot be identified visually.

In keratin electrophoresis, feather proteins from an unknown sample are compared with samples from known specimens stored in databases.

9.4.4 DNA Analysis

Small amounts of blood or tissue may be used for DNA analysis, comparing bird strike material with genetic library sequences shows that a 97–99% efficiency in matching is possible. However, DNA analysis may cost US$ 15 000 for each sample.

DNA barcoding is the newest tool available for bird strike. It involves extracting the mtDNA for the cytochrome c oxidase subunit 1, COI or COX1 gene, known as the "barcode gene". This is extracted from bird strike samples that consist of blood and tissue and then the unknown sequence is matched to a DNA library available on the Barcode of Life Database (BoLD) [13, 17].

An analysis for bird strike remains during the fall (autumn) migration of 2006 (between 1 September and 31 December) collected by the USAF and the FAA was performed using DNA at the Smithsonian Institution [13]. The aim was to evaluate current collecting methods for small amounts of bird strike evidence, consisting of only blood or tissue, and to analyze the condition of the samples submitted for DNA testing.

The success of DNA analyses depends on the age of the sample. A detailed study for such a parameter was carried out by Dove et al. [13]. The findings outlined that the amount of time from the bird strike event to DNA sampling depends on the week (Table 9.1) [13].

Table 9.1 Dependence of success of DNA methods on the age of the sample.

Week	Days	% of DNA success
1	1–7	52.6
2	8–14	67.1
3	15–21	66.1
4	22–28	76.9
+	29–116	72.3

Samples that were received more than 29 days after the event resulted in viable DNA in 72.29% of the cases. The highest success rate was found in samples that were received from 22 to 28 days after the strike (76.9%). The age of the sample was not critical to DNA extraction success indicating that bird strike evidence from events far in the past can be identified using DNA methods [13]. However, this conclusion from the author's point of view needs more research for verification.

The poor condition of the samples collected in the fall migration (1 September to 31 December 2006) led to the following new recommendations relating to current field collecting techniques for all bird strike evidence consisting of blood and tissue:

1. the use of alcohol (ethanol or isopropyl alcohol) or alcohol-based towelettes, instead of water and paper towels, to wipe the bird evidence from the aircraft , or
2. use of a DNA "fixing" card such as the Whatman FTA card to prevent DNA degradation as soon as possible.

9.5 Accident Analysis

On 4 March 2008, a Cessna Citation 1 (Model 500) encountered a flock of birds after takeoff and crashed nearly 7 km from Wiley Post Airport, Oklahoma, killing all five people on board [3].

Bird remains were found on the horizontal and vertical stabilizer and the right interior cowling. Both microscopic and DNA samples were submitted from the field samples. Later, items from the cockpit, windshield, and engine were examined in the Smithsonian Feather Identification Lab.

A single portion of a small white feather and seven feather barbs (three downy barbs and four pennaceous barbs) were found. Microscopic investigation outlined the similarity of the three downy barbs. They included unpigmented barbs and short barbules with long prongs on the distal portion of the barbule (Figure 9.7).

These microscopic features are typical of several birds that could occur in Oklahoma, including loons, grebes, white pelicans, and cormorants. Because neither the visual inspection nor the microscopic findings could identify the exact bird species, it was decided to rely on the DNA analysis.

After more than a hundred tests, it was confirmed that the bird that had hit the aircraft was an American white pelican. The microscopic structures of the downy barbules were reexamined and it was confirmed that the feather characteristics were consistent

Figure 9.7 Photomicrograph of a downy feather collected from the tail section of the Cessna Citation [3]. *Source:* reproduced with permission.

with those of an American white pelican. This finding matches with the fact that the American white pelican is common during the spring in Oklahoma.

References

1 International Birdstrike Committee (2006). Recommended Practices No. 1: Standards For Aerodrome Bird/Wildlife Control. http://www.int-birdstrike.org/Standards_for_Aerodrome_bird_wildlife%20control.pdf (accessed 12 January 2019).

2 Smithsonian Institution, Feather Identification Lab. https://www.faa.gov/airports/airport_safety/wildlife/smithsonian (accessed 12 January 2019).

3 Dove, C.J., Dahlan, N.F., and Heacker, M. (2009). Forensic bird-strike identification techniques used in an accident investigation at Wiley Post Airport, Oklahoma, 2008. *Human-Wildlife Interactions* 3 (2): article 6. https://digitalcommons.usu.edu/hwi/vol3/iss2/6/ (accessed 12 January 2019).

4 Brom, T.G. (1991). *The Diagnostic and Phylogenetic Significance of Feather Structures*. Amsterdam: Instituut Voort Taxomomische Zoologische.

5 Dove, C.J. (2000). *A Descriptive And Phylogenetic Analysis of Plumulaceous Feather Characters in Charadriiformes*. McLean, VA: American Ornithologists' Union.

6 Dove, C.J. and Agreda, A. (2007). Differences in plumulaceous feather characters of dabbling and diving ducks. *Condor* 109: 192–199.

7 Rajaram, A. (2002). Barbule structure of bird feathers. *Journal of Bombay Natural History Society* 99: 250–257.

8 Dove, C.J., Rotzel, N.C., Heacker, M., and Weigt, L.A. (2008). Using DNA barcodes to identify species involved in bird strikes. *Journal of Wildlife Management* 72: 1231–1236.

9 Smithsonian Institution, Feather Identification Lab. https://www.faa.gov/airports/ airport_safety/Wildlife/smithsonian/ (accessed 15 January 2019).

10 Brom, T.G. (1984). Microscopic identification of feathers in order to improve birdstrike statistics. Proc. Conf. Wildlife Hazards to Aircraft (Charleston, SC), report no. DOT/ FAA/AAS/84-1: pp. 107–120.

11 Ouellet, H. and van Zyll de Jong, S.A. (1990). Feather Identification by Means of Keratin Protein Electrophoresis. Bird Strike Committee Europe, Working Papers, 20th Meeting, Helsinki. http://www.int-birdstrike.org/Helsinki_Papers/IBSC20%20WP8.pdf (accessed 15 January 2019).

12 Mathur, U.B. and Adaval, S.K., Laboratory studies on avian blood under simulated crash conditions, *Indian Journal of Aerospace Medicine* 2007, pp 51–54. http://medind. nic.in/iab/t07/s1/iabt07s1p51.pdf (accessed 12 January 2019).

13 Dove, C.J., Heacker, M., and Rotzel, N. (2007). The Birdstrike Identification Program at the Smithsonian Institution and New Recommendations for DNA Sampling. 2007 Bird Strike Committee USA/Canada, 9th Annual Meeting, Kingston, Ontario. https://pdfs. semanticscholar.org/4f3e/6163331a6020518f18bb6af3cf678f3d4b5f.pdf (accessed 12 Janaury 2019).

14 Taxonomic Rank. https://en.wikipedia.org/wiki/Taxonomic_rank (accessed 15 January 2019).

15 What is a "species"? http://research.amnh.org/vz/ornithology/crossbills/species.html (accessed 15 January 2019).

16 Dolbeer, R.A. and Wright, S.E. (2009). Safety management systems: how useful will the FAA National Wildlife Strike Database be? *Human–Wildlife Conflicts* 3: 167–178.

17 Gaikwad, S., Munot, H., and Shouche, Y. (2016). Utility of DNA barcoding for identification of bird-strike samples from India. *Current Science* 110 (1): 10. https:// www.researchgate.net/publication/309634391_Utility_of_DNA_barcoding_for_ identification_of_bird-strike_samples_from_India (accessed 15 January 2019).

Index

Bird Strike in Aviation: Statistics, Analysis and Management, First Edition. Ahmed F. El-Sayed.
© 2019 John Wiley & Sons Ltd. Published 2019 by John Wiley & Sons Ltd.